Few-Body Systems

Suppl. 3

Few-Body Systems

Editor-in-Chief: H. Mitter, Graz
Associate Editor: W. Plessas, Graz

Supplementum 3

H. Arenhövel and M. Sanzone

Photodisintegration of the Deuteron

A Review of Theory and Experiment

Springer-Verlag Wien New York

H. Arenhövel
Institut für Kernphysik, Johannes-Gutenberg-Universität, Mainz,
Federal Republic of Germany

M. Sanzone
Dipartimento di Fisica, Università di Genova, and Istituto Nazionale di Fisica Nucleare,
Sezione di Genova, Genova, Italy

With 94 Figures

Printed in Austria by Novographic, Ing. W. Schmid, A-1238 Wien

Printed on acid-free paper

ISSN 0177-8811
ISBN 3-211-82276-3 Springer-Verlag Wien – New York
ISBN 0-387-82276-3 Springer-Verlag New York – Wien

Acknowledgement

We would like to thank P. Bosted, P. Doll, K.H. Krause, J. Matthews, and R.O. Owens for sending us experimental data, M. Anghinolfi for the evaluation of the coefficients A_n of some experimental angular distributions, and J.L. Friar, W. Jaus, and B. Mosconi for sending us theoretical results. Special thanks go to H. Göller for discussions on relativistic effects, to K.-M. Schmitt and P. Wilhelm for their help in numerous calculations, to J. Ahrens for critical reading of section 5 and to K.-M. Schmitt for critical reading of the whole manuscript. Last but not least we are very grateful to Heidrun Tuba and Elisabeth Murtinger for their excellent typing and their most efficient and untiring help in preparing this manuscript.

Table of Contents

1 Introduction

More than 50 years ago, in 1934, Chadwick and Goldhaber (ChG 34) published a paper entitled "A 'Nuclear Photo-effect': Disintegration of the Diplon by γ-Rays."[1] They noted in the introduction:

> "By analogy with the excitation and ionisation of atoms by light, one might expect that any complex nucleus should be excited or 'ionised', that is, disintegrated, by γ-rays of suitable energy",

and furthermore:

> "Heavy hydrogen was chosen as the element first to be examined, because the diplon has a small mass defect and also because it is the simplest of all nuclear systems and its properties are as important in nuclear theory as the hydrogen is in atomic theory".

Almost at the same time, in 1935, the first theoretical paper on the photodisintegration of the deuteron entitled "Quantum theory of the diplon" by Bethe and Peierls (BeP 35) appeared. It is not without significance that these two papers mark the beginning of photonuclear physics in general and emphasize in particular the special role the two-body system has played in nuclear physics since then and still plays. A steady flow of experimental and theoretical papers on deuteron photodisintegration and its inverse reaction, n-p capture, shows the continuing interest in this fundamental process (see fig. 1.1).

In fact, the study of N-N scattering and the bound deuteron system as well has furnished us with a wealth of detailed information on the elementary N-N interaction allowing the construction of various realistic N-N potential models, more or less phenomenological with input from meson theories and dispersion methods. On the other hand, one has no direct information on the internal spatial structure of the two-body system at small and intermediate distances from such scattering experiments, since only the asymptotic parts of the wave functions via the phase shifts determine the experimental observables like cross sections and polarization asymmetries.

Therefore, in order to study the interior part of the two-body wave function one needs a suitable probe. It is well known that a very sensitive and clean probe is given by the electromagnetic interaction in photon absorption and electron scattering, because the interaction is well understood and sufficiently weak in order to allow a lowest-order perturbative approach. This then means that the comparison between theory and experiment is an important and direct test for the underlying theoretical framework and any significant discrepancy will point to shortcomings of the theoretical description such as approximations in the actual calculation or failure of the basic assumptions.

From the experimental point of view, as this process leads to a break-up of the nucleus in two particles only and since no other reaction channels are possible, at least as long as the excitation energy is below the pion production threshold, the measurement of

[1] A personal account of the history of this pioneering experiment has recently been given by M. Goldhaber (AIP Conf. Proc. 185, S.-U. Chung, ed. (1989)).

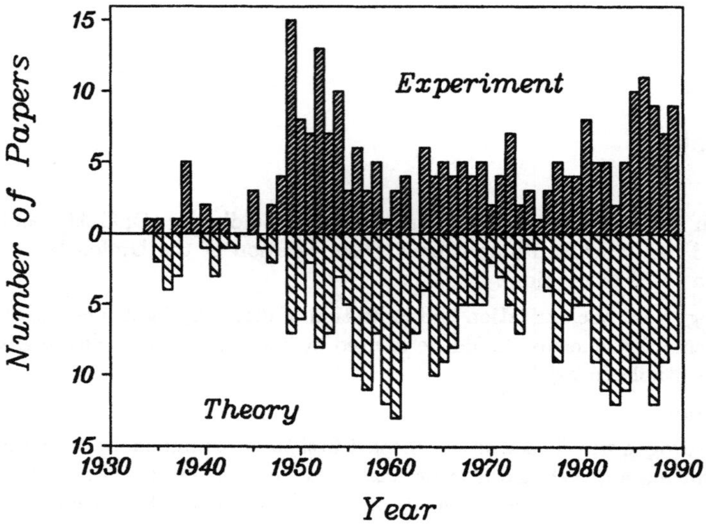

Fig. 1.1: Histogram of experimental and theoretical papers per year on deuteron photodisintegration and radiative *n-p* capture.

the yield of protons emitted is a direct measurement of the deuteron disintegration cross section. Moreover, since the deuteron has no bound excited states, the measurement of the proton energy and direction is also a measurement of the energy of the incoming probe. This is very important when a continuous spectrum is used as in the case of a bremsstrahlung photon beam.

Another remark with respect to the experimental conditions concerns the binding energy: apart from ^{9}Be for the break-up in $2\alpha + n$, ^{2}H is the only stable nucleus with a small binding energy. For this reason deuteron photodisintegration allows to study transitions to highly excited continuum states of the *n-p* system by using available gamma sources from nuclear reactions.

It is thus evident why a large part of experiments in photonuclear physics has been devoted to the photodisintegration of the deuteron. It is the simplest process to be investigated experimentally and even if a deuterium target is not a simple one, many experiments have been performed using almost all kind of accelerators and techniques. The special role of the deuteron is also illustrated by the fact that everytime when new experimental techniques and/or new generations of accelerators have been developed and everytime when new theoretical concepts on the nature of the *N-N* interaction have been introduced like, e.g., meson-exchange models and isobar configurations or most recently quark-gluon degrees of freedom, the deuteron has served as a test case.

Different aspects of deuteron photodisintegration have been reviewed by various authors. The early history is described in Fel 55, BiW 57, HuS 57 and Lev 60. More recent reviews on the conventional non-relativistic theory and on subnuclear degrees of freedom can be found in Bre 73, Bai 79, Are 85 and Are 86 and on the experimental achievements in Cam 84, Sor 85 and San 86. The possible evidence for the existence of dibaryon resonances is reviewed in Sch 81, Loc 84, Mak 84, LoS 86 and Mar 88.

In view of the new concepts in the theory of strong interactions it is timely to assess the quality of presently available experimental data and the status of our understanding of this important process in the framework of present theoretical concepts. Because possible evidence for explicit quark-gluon dynamics and related more exotic phenomena like dibaryon resonances can be obtained most likely only if the present conventional theory with nucleon, meson and isobar degrees of freedom is pushed to its limits. For this reason we will try to give a more comprehensive review on deuteron photodisintegration in the hope that this will enable us to see in which directions we might encounter the limits of present concepts. Certainly, this review will not be exhaustive and complete. The vast material had to be selected and filtered. It is obvious that such a procedure leads unavoidably to choices subject to personal preferences and judgements. Therefore, we apologize here to all who feel that their work has not been appropriately evaluated or referred to.

After a brief outline of the early elementary theory and the historical development since then in section 2 we shall consider the general formalism in section 3. Kinematical quantities, the density matrices for the description of polarization, the T-matrix and the general expressions for differential cross section and polarization observables will be discussed there. Section 4 is devoted to a survey of the various theoretical methods and concepts for the calculation of the T-matrix. The different beams for photodisintegration and capture experiments are described in section 5 while in section 6 present experimental data on total and differential cross sections and on polarization observables as well are discussed. In section 7 we shall present theoretical results and discuss how well we understand the existing experimental data within the present theoretical concepts. Finally, we shall give some conclusions and an outlook in section 8.

2 Elementary theory and brief history

In the history of the theory of deuteron photodisintegration one may distinguish roughly three periods:

(i) the **primitive period** of the elementary theory using very simple wave functions and forces and considering lowest multipoles ($E1, M1$) only,

(ii) the **classical period** still in the framework of conventional nuclear physics, but using realistic forces with correspondingly elaborate wave functions and considering also higher multipole transitions,

(iii) the **post-classic period** with explicit treatment of subnuclear degrees of freedom like meson and isobar degrees of freedom and very recently quark-gluon degrees of freedom.

The elementary approach of Bethe and Peierls (BeP 35) considers dipole transitions only in the long-wave-length limit and uses very simple wave functions for the bound and continuum states. In dipole approximation the differential absorption cross section of a photon with energy ω is given by

$$\frac{d\sigma}{d\Omega} = 2\pi^2 e^2\, \omega k M |\boldsymbol{D}_{fi} \cdot \boldsymbol{\varepsilon}|^2 \,, \tag{2.1}$$

where $\boldsymbol{D}_{fi} = \langle f|\hat{\boldsymbol{D}}|i\rangle$ denotes the dipole transition matrix element between the deuteron bound state and the continuum and $\boldsymbol{\varepsilon}$ the photon polarization vector. The average nucleon mass is denoted by M and the outgoing nucleon momentum by $\boldsymbol{k}$. For the N-N force the simplest ansatz has been taken, namely a spin-independent zero-range δ-force

$$V(\boldsymbol{r}) = -V_0\, \delta(r) \,. \tag{2.2}$$

The justification for this is that in the long-wave-length limit the photon wave length is large compared to the range of the potential so that details of the force are not expected to matter. This force acts only in S-states and one obtains for the bound state solution

$$\psi_s^0(\boldsymbol{r}) = \frac{1}{\sqrt{4\pi}} \frac{u_s^0(r)}{r} \tag{2.3}$$

with

$$u_s^0(r) = N_s^0 \mathrm{e}^{-\alpha r} \,, \tag{2.4}$$

where α is related to the deuteron binding energy ϵ by

$$\alpha = \sqrt{M\epsilon} \,. \tag{2.5}$$

The normalization constant of the radial wave function is given by

$$N_s^0 = \sqrt{2\alpha} \,. \tag{2.6}$$

Since the dipole operator does not allow transitions into S-states one can insert a free plane wave as final state and obtains

$$D_{fi} = \frac{N_s^0}{2(2\pi)^{3/2}} \int d^3 r e^{-i\boldsymbol{k}\cdot\boldsymbol{r}} \boldsymbol{r} \psi_s^0(\boldsymbol{r}) , \tag{2.7}$$

where $\boldsymbol{k}$ is the final-state relative momentum $\boldsymbol{k} = (\boldsymbol{p}_p - \boldsymbol{p}_n)/2$ related to the photon energy ω by

$$k = \sqrt{M(\omega - \epsilon)} \quad \text{or} \quad a^2 + k^2 = M\omega \tag{2.8}$$

neglecting the recoil energy. The integration yields

$$D_{fi} = i\frac{\sqrt{a}}{\pi} \frac{\boldsymbol{k}}{(\alpha^2 + k^2)^2} . \tag{2.9}$$

This then gives for the differential cross section averaged over the photon polarizations

$$\frac{d\sigma(E1)}{d\Omega} = \frac{e^2}{\alpha^2} \frac{(\gamma - 1)^{3/2}}{\gamma^3} \sin^2 \Theta, \quad \gamma = \frac{\omega}{\epsilon} , \tag{2.10}$$

where Θ is the angle between incident photon and proton momentum. We would like to remark that the same result would have been obtained, if one had used the electromagnetic interaction in the form $-\boldsymbol{j} \cdot \boldsymbol{A}$, where $\boldsymbol{j}$ denotes the nuclear one-body current and $\boldsymbol{A}$ the vector potential of the electromagnetic field. This means the absence of exchange current effects.

For the total cross section one obtains from (2.10) the so-called Bethe-Peierls cross section (BeP 35)

$$\sigma_{BP}(\omega) = \frac{8\pi}{3} \frac{e^2}{\alpha^2} \frac{(\gamma - 1)^{3/2}}{\gamma^3} . \tag{2.11}$$

The characteristic feature of (2.10) is a pure $\sin^2 \Theta$ dependence of the angular distribution of the emitted proton, which is related to the dipole approximation in conjunction with the pure S-state of the deuteron. The total cross section shows a maximum at twice the binding energy and then falls off like $\omega^{-3/2}$. The integrated cross section fulfills the classical TRK-sum rule (Lev 60) as it should

$$\sigma_{int}(E_1) = \int_\epsilon^\infty d\omega \, \sigma_{BP}(\omega) = \pi^2 \frac{e^2}{M} . \tag{2.12}$$

Comparison with experiment shows that the shape of the total cross section is quite well reproduced even at energies where one would not expect it in view of the underlying simple assumptions. However, the absolute size is missed by a factor of about 1.7, i.e., the theory is much too low.

This underestimation of the cross section is related to the finite size of the NN interaction as has first been noted by Bethe and Longmire (BeL 50). The use of the zero range wave function has been justified by the observation that the photon wave length is much larger than the range of the nuclear force, so that for the evaluation of the dipole matrix element the asymptotic form of the wave function is sufficient. This argument is valid concerning the relative radial dependence but not for the absolute

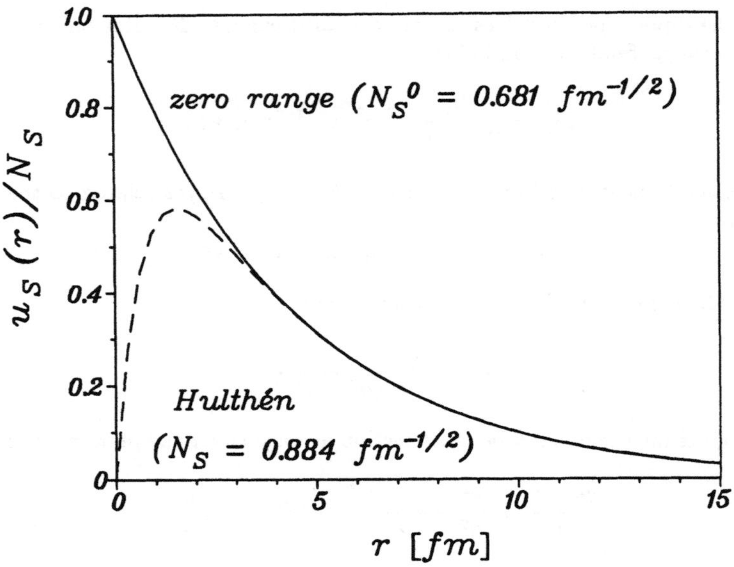

Fig. 2.1: Comparison of a zero range with a finite-range bound S-state wave function

value of the asymptotic amplitude. That is easily seen if one compares a finite size wave function u_s with a zero range one u_s^0, as is demonstrated in fig. 2.1.

Taking out the asymptotic normalization, i.e., writing

$$u_s(r) \underset{r\to\infty}{\longrightarrow} N_s\, e^{-\alpha r}\,, \tag{2.13}$$

then the normalization condition gives

$$\int_0^\infty dr \left(\left(\frac{u_s^0}{N_s^0}\right)^2 - \left(\frac{u_s}{N_s}\right)^2\right) = \frac{1}{(N_s^0)^2} - \frac{1}{(N_s)^2}\,. \tag{2.14}$$

Looking at fig. 2.1, one readily notes that the difference is positive for a finite-size wave function, and, therefore, $N_s > N_s^0$, i.e., the asymptotic amplitude of a finite-size wave function is larger than that of the zero-range one.

Bethe and Longmire have expressed the difference using the effective range theory in terms of the triplet effective range r_t in the shape-independent approximation

$$\int_0^\infty dr \left(\left(\frac{u_s^0}{N_s^0}\right)^2 - \left(\frac{u_s}{N_s}\right)^2\right) = \frac{1}{2}r_t\,. \tag{2.15}$$

Then, one obtains

$$\frac{1}{N_s^2} = \frac{1}{2\alpha}(1-\alpha r_t) \tag{2.16}$$

and thus

$$N_s = \sqrt{\frac{2\alpha}{1-\alpha r_t}}\,. \tag{2.17}$$

Taking this asymptotic normalization into account means to multiply the Bethe-Peierls cross section of (2.11) by $(N_s/N_s^0)^2$, i.e., by $(1-\alpha r_t)^{-1}$. Then, one obtains the so-called effective range cross section (BeL 50)

$$\sigma_{ER} = \frac{\sigma_{BP}}{1-\alpha r_t} = \frac{8\pi}{3}\frac{e^2}{\alpha^2}\frac{(\gamma-1)^{3/2}}{\gamma^3(1-\alpha r_t)} . \tag{2.18}$$

Remarkably, this expression contains experimental observables only and, therefore, is independent of any potential model. Using the experimental value $r_t = 1.759 \pm 0.005$ fm (DuK 83), one obtains surprisingly good agreement with experimental data which have been selected in section 6.3 (see fig. 2.2). Only very close to the threshold one finds larger deviations as becomes apparent in fig. 2.3, where the data are normalized to σ_{ER}. However, the seemingly good agreement of this most simple theory with experimental data should not lead one to the conclusion that deuteron photodisintegration is an uninteresting subject. On the contrary, it will become clearer when we discuss further details and in particular angular distributions and polarization observables that this process is a fascinating subject allowing deeper insights into the nature of strong interaction physics.

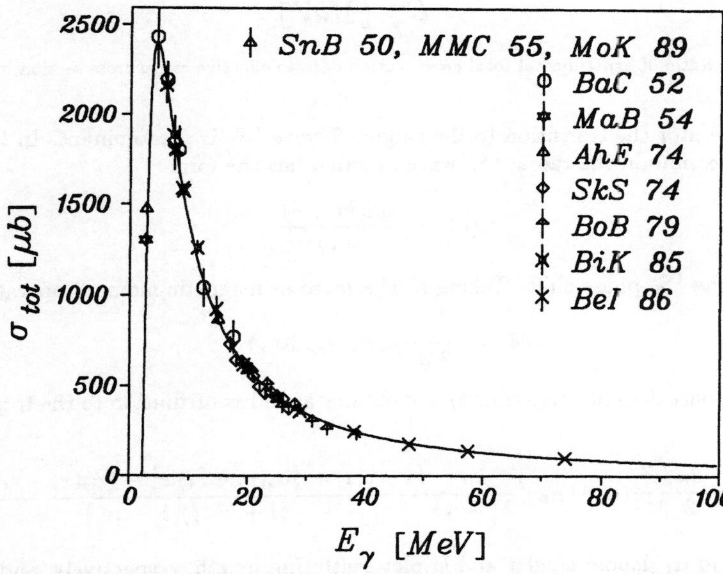

Fig. 2.2: Comparison of the effective range cross section σ_{ER} with selected experimental data as discussed in section 6.3.

The deviation near threshold in fig. 2.3 is due to the neglect of $M1$ contributions, which are appreciable there. This has first been noted by Fermi (Fer 35) who has calculated the $M1$ contribution to slow n-p capture and photodisintegration near threshold. He was motivated by the experimental observation of slow neutron capture in paraffin while the Bethe-Peierls theory predicted a vanishing capture rate. Later Bethe and Longmire (BeL 50) have evaluated the $M1$ cross section under similar simple assumptions than for E1 by using the asymptotic forms of the effective range theory for the initial and final state wave functions. Due to the dominance of the isovector magnetic

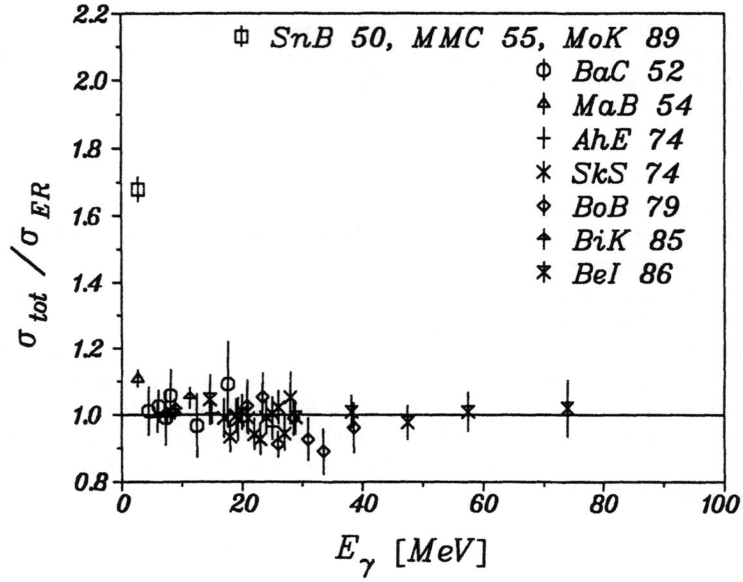

Fig. 2.3: Ratio of experimental total cross section data to effective range cross section σ_{ER}.

moment operator the transition to the singlet S-wave ${}^1S_0$ is predominant. In the zero range approximation the radial ${}^1S_0$ wave function has the form

$$u_s(r) = \frac{\sin(kr+\delta)}{\sin\delta}, \tag{2.19}$$

where δ is the ${}^1S_0$ phase shift. Taking furthermore as magnetic moment operator

$$M = \frac{e}{2M}(\mu_p\sigma_p + \mu_n\sigma_n), \tag{2.20}$$

(the orbital part does not contribute) one obtains as $M1$ contribution to the total cross section

$$\sigma_{ER}(M1) = \frac{2\pi}{3}\frac{e^2}{M^2}(\mu_p-\mu_n)^2\frac{k\alpha}{k^2+\alpha^2}\frac{\left(1-\alpha a_s+\frac{1}{4}a_s(r_s+r_t)\alpha^2-\frac{1}{4}a_s(r_s-r_t)k^2\right)^2}{(1+k^2a_s^2)(1-\alpha r_t)}, \tag{2.21}$$

where α_s and α_t denote singlet and triplet scattering length, respectively, and r_s the singlet effective range. The last factor corrects finite size effects (BeL 50, Aus 53). Shape-dependent corrections and D-state effects as well have been neglected. They are discussed in Noy 65. This cross section is shown in fig. 2.4 together with $\sigma_{ER}(E1)$ for comparison.

The virtue of this formula is that it is again independent of any details of the nuclear potential and that it contains only measurable quantities. The only critical assumption is the absence of exchange contributions to the magnetic-moment operator. In fact, such exchange effects have been invoked for a long time to account for the discrepancy between the theory according to (2.21) and experiment (CoW 65) giving a 10 percent higher value than (2.21) (Aus 53, AuR 60, Noy 65) as will be discussed later in section 7.1.

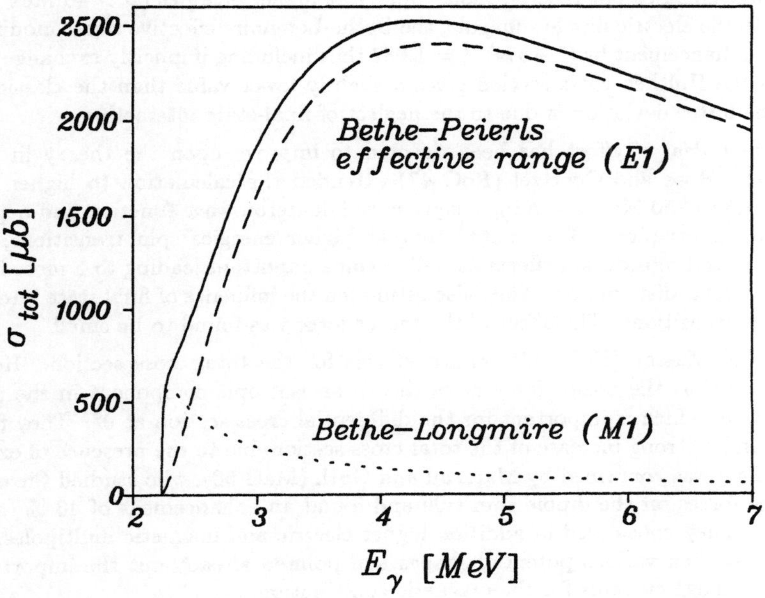

Fig. 2.4: Electric and magnetic dipole total cross sections in the effective range theory and the sum (full curve).

At the same time, when the effective range theory has been developed, finite size effects of the deuteron wave function have been considered explicitly by various authors using a Hulthén wave function for the deuteron bound state S-wave

$$u_s^H(r) = \sqrt{\frac{2\alpha}{1-\alpha r_t}}(e^{-\alpha r} - e^{-\beta r}) \tag{2.22}$$

as solution for the Hulthén potential

$$V_H(r) = -\frac{1}{M}(\beta^2 - \alpha^2)\frac{e^{-(\beta-\alpha)r}}{1-e^{-(\beta-\alpha)r}} \, . \tag{2.23}$$

Through the normalization condition β is related to the effective range

$$r_t = \frac{4}{\alpha+\beta} - \frac{1}{\beta} \, , \tag{2.24}$$

giving $\beta = 1.378$ fm $\approx 6\alpha$. Levinger (Lev 49), Schiff (Sch 50) and Marshall and Guth (MaG 50) arrived then at the following expression for the total cross section

$$\sigma_H(\omega) = \left(1 - \frac{\gamma^2}{\left(\gamma + \left(\frac{\beta}{\alpha}\right)^2 - 1\right)^2}\right)^2 \sigma_{ER}(\omega) \, . \tag{2.25}$$

One readily sees that this expression coincides with the Bethe-Longmire cross section at low energies but falls off more rapidly at higher energies ($\omega > 40$ MeV). With respect to the electric dipole sum rule, the Bethe-Longmire effective range modification gives an enhancement by $(1-\alpha r_t)^{-1} = 1.687$ thus including implicitly exchange effects, whereas the Hulthén cross section gives a slightly lower value than the classical sum rule. The latter deviation is due to the neglect of final-state interaction.

A great deal of effort has been devoted to improve upon the theory in various directions. Rose and Goertzel (RoG 47) extended the calculation to higher photon energies up to 250 MeV assuming a square-well deuteron wave function and neglecting final-state interaction. They noted that at higher energies spin transitions, higher multipoles and retardation effects as well become important leading to a more forward peaked angular distribution. They also estimated the influence of final state interaction for dipole transitions. The effect of the tensor force was found to be small.

Hu and Massey (HuM 49) confirmed this for the total cross section. However, they noted that the tensor force gives rise to an isotropic component in the angular distribution, which is important for the differential cross section at 0°. They furthermore found a strong increase of the total cross section due to the presence of exchange forces. This was confirmed by Marshall and Guth (MaG 50), who studied the effect of exchange forces on the dipole sum rule and found an enhancement of 40 % (see also LeB 50). They considered in addition higher electric and magnetic multipoles. Schiff (Sch 50) studied various potential shapes and pointed already out the importance of meson-exchange currents for the magnetic contribution.

A deuteron wave function including a D-wave has been used by Feshbach and Schwinger (FeS 51) by solving numerically the coupled $S-D$ equations with a tensor force of Yukawa type. These authors note also that, by taking for the electromagnetic interaction $e\boldsymbol{D}\cdot\boldsymbol{E}$ instead of $\boldsymbol{j}_{(1)}\cdot\boldsymbol{A}$ with the one body-current $\boldsymbol{j}_{(1)}$, one has implicitly included contributions from exchange currents. Berger (Ber 54) considered all multipoles without resorting to a systematic multipole expansion by neglecting the final-state interaction except for the S-states. He studied various effects of zero- and finite-range S-wave functions, central and non-central forces. He again pointed out that the major effect of higher multipoles is a change of the shape of the angular distribution at higher energies.

The beginning of the classical period is marked by the work of de Swart and Marshak (DSM 58, 59, DSw 59) using more realistic potentials but still restricting themselves to dipole transitions only. They made a detailed analysis of the various final-state contributions. In particular, they noted again the large effect of the tensor force on the isotropic part of the angular distribution and concluded that a large D-wave probability P_D is necessary to account for the experimental value. We would like to remark already here that this conclusion has recently been reversed when higher multipoles have been shown to contribute appreciably to the isotropic part giving now an isotropic part too high with respect to the experiment (see section 7.3). Similar calculations have been done by Hsieh (Hsi 56a-b, 59a-c) and by Nicholson and Brown (NiB 59) stressing again the importance of all possible transitions. Furthermore, they state that retardation in the multipoles gives only a small effect at $E_\gamma = 130$ MeV in contrast to the results of Matsumoto (Mat 60, 63) and Donnachie and O'Donnell (DoO 64a), who found considerable retardation effects even at lower energies.

The sensitivity of the polarization of the outgoing nucleons on magnetic transitions due to $M1-E1$ interference has been emphasized by various authors (Ros 56, CzS 57,

58a-b, Kaw 58, DSC 59, KrM 60, Saw 64). Since $E1$ transitions dominate the angular distributions, the nucleon polarization allows to study magnetic transitions in order to shed light on the spin-dependent part of the N-N interaction. Furthermore, the role of possible mesonic contributions in magnetic transitions has been studied by Kramer and Müller (KrM 60) for low energies ($\leq$ 11 MeV). They conclude that nucleon polarization should be more sensitive to meson-exchange currents than angular distributions at these low energies because magnetic transitions are more important here due to the above-mentioned interference.

A certain endpoint of this classical period has been reached by Rustgi et al. (RuZ 60), Donnachie and O'Donnell (Don 62a, DoO 64a-b) and Partovi (Par 64) in detailed calculations within the framework of classical non-relativistic nuclear theory but without taking into account explicit exchange currents. However, by using the Siegert operators for electric transitions they included the major portion of meson-exchange currents implicitly as will be discussed in section 4.1. They have studied a variety of approximations, the importance of higher multipoles, and included photon and nucleon polarization effects.

At the same time, in the early 60's, covariant theories have been developed by Pearlstein and Klein (PeK 60) using an S-matrix approach and by Sakita and Goebel (SaG 62) and Donnachie (Don 62b) using dispersion relations. Pearlstein and Klein were aiming at a consistent description of meson degrees of freedom in order to obtain a unified treatment of the low energy region up to photon energies of about 100 MeV and the high energy region of the Δ resonance, because the latter had been considered in rough models only (Aus 55, Wil 56, Zac 56).

In the dispersion theoretical approach the main idea was to have a theoretical framework independent of an underlying non-relativistic potential model which allows in principle to incorporate directly the N-N phase shifts. It has been further developed by Le Bellac, Renard and Tran Thanh Van (LeR 64a-b) and applied to the region below π-threshold. More recently, it has been extended to the high-energy domain by various authors (see section 4.5).

In the meantime the main interest has shifted from the study of the N-N interaction in terms of potential models to the study of subnuclear degrees of freedom. In fact, the manifestation of such degrees of freedom in meson-exchange effects and Δ excitation has revived the interest in deuteron photo- and electrodisintegration and its inverse, the radiative n-p capture. One particular aspect at higher energies has been the still open question of possible evidence of dibaryon resonances which has stimulated quite a few experimental and theoretical studies. Furthermore, with the development of quantum chromodynamics the possible role of quark-gluon degrees of freedom is now receiving increasing attention.

3 Formal theory for two-particle channels

The general formalism has been developed over many years by various authors. One starting point is the work of de Swart (DSw 59) who has considered electric multipoles in the long-wave-length limit using the Siegert theorem and as magnetic contribution only the dipole spin-flip transition. The T-matrix is then expanded in terms of reduced multipole amplitudes. This approach has been generalized by Donnachie (Don 62a) and Partovi (Par 64) by including higher electric and magnetic multipoles. Furthermore, the electric multipoles are not restricted to the long-wave-length limit and the additional terms besides the Siegert operators (see section 4.1) are included. Using techniques from angular momentum algebra Cambi, Mosconi and Ricci (CaM 82a) have considerably simplified the expressions for differential cross section, photon asymmetry and nucleon polarization expressing them in terms of Legendre polynomials or functions, respectively. That has subsequently been generalized to all polarization observables in Are 88a.

A different approach has been followed by Rustgi, Zernik, Breit and Andrews (RuZ 60) and further developed in BoR 79 and ZiR 79 by expanding the T-matrix in terms of singlet and triplet spin-angular functions. As pointed out in FrG 84, these can be expressed in terms of an irreducible set of twelve spin invariants which form the non-relativistic analog of the twelve invariants of the covariant approach (SaG 62, Don 62b, LeR 64a), which constitutes the third line of the formal development. In the following we will mainly consider the de Swart-Partovi formalism because it is the most elaborate one and has been widely used.

We now will describe the general formalism of deuteron photodisintegration limiting ourselves to pure two-body break-up, i.e., two nucleons only in the final states. That means, we will not consider explicit π-channels even above π-production threshold.

3.1 Kinematics

The initial state is characterized by photon and deuteron four-momenta, p_γ and p_d (see fig. 3.1.1), and by density matrices for photon polarization and deuteron orientation ρ^γ and ρ^d, respectively. Correspondingly, the final state is described by the four momenta of proton and neutron, p_p and p_n, respectively, and density matrices ρ^p and ρ^n, if a polarization analysis is performed.

We shall consider two reference frames: the lab frame and the cm frame. The latter will be our preferred reference frame and all quantities without further specification will be referred to it. Lab frame quantities will be indicated by a superscript "L".

First we will give some kinematical relations between lab and cm frame quantities. Denoting in the lab frame

$$p_\gamma^L = (\omega^L, \omega^L) \tag{3.1.1}$$

$$p_d^L = (M_d, 0) \tag{3.1.2}$$

and in the cm system

$$p_\gamma = (\omega, \boldsymbol{\omega}) \tag{3.1.3}$$

$$\begin{aligned} p_d &= (E_d, -\boldsymbol{\omega}) & E_d &= (M_d^2 + \omega^2)^{1/2} \quad (3.1.4) \\ p_p &= (E_p, \boldsymbol{k}) & E_p &= (M_p^2 + \boldsymbol{k}^2)^{1/2} \\ &= p_1 & & \quad (3.1.5) \\ p_n &= (E_n, -\boldsymbol{k}) & E_n &= (M_n^2 + \boldsymbol{k}^2)^{1/2} \\ &= p_2 & & \quad (3.1.6) \end{aligned}$$

and introducing the total cm energy W, which is the invariant mass,

$$W \;=\; E_p + E_n \;=\; \omega + E_d \tag{3.1.7}$$

and the asymptotic final-state kinetic energy

$$E_{np} \;=\; W - M_p - M_n \tag{3.1.8}$$

one has the following relations between cm and lab frame energies

$$\omega \;=\; \omega^L \left(1 + \frac{2\,\omega^L}{M_d}\right)^{-1/2} \tag{3.1.9}$$

$$W \;=\; M_d \left(1 + \frac{2\,\omega^L}{M_d}\right)^{1/2} . \tag{3.1.10}$$

Furthermore, one obtains using energy conservation

$$\boldsymbol{k}^2 = \frac{1}{4W^2}(W^2 - (M_p + M_n)^2)(W^2 - (M_p - M_n)^2) \,, \tag{3.1.11}$$

which reduces for $M_p = M_n = M$ to

$$\boldsymbol{k}^2 \;=\; \frac{1}{4}(W^2 - 4M^2) \;=\; \frac{1}{4}E_{np}(E_{np} + 4M) \,. \tag{3.1.12}$$

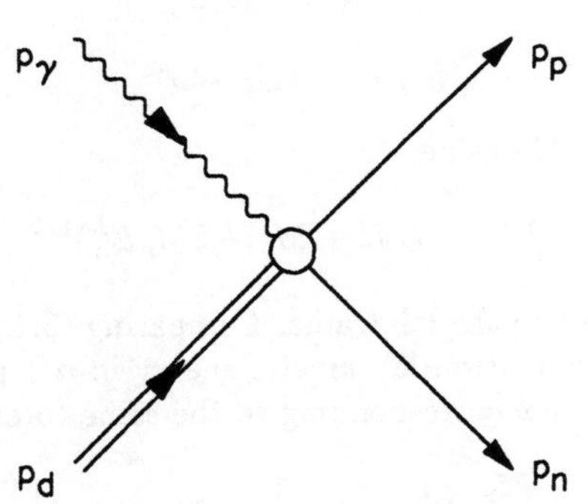

Fig. 3.1.1: Diagram for deuteron photodisintegration with definition of momenta

It is useful to introduce total and relative momentum for the two-particle channels by

$$P = p_1 + p_2 \tag{3.1.13}$$

$$p = \frac{1}{2}(p_1 - p_2)\,. \tag{3.1.14}$$

Then, one has for the initial and final states in the cm frame

$$P_i = (W, \boldsymbol{0}) \tag{3.1.15}$$

$$p_i = \left(\frac{1}{2}(\omega - E_d), \boldsymbol{\omega}\right) \tag{3.1.16}$$

$$P_f = P_i \tag{3.1.17}$$

$$p_f = \left(\frac{1}{2}(E_p - E_n), \boldsymbol{k}\right)\,. \tag{3.1.18}$$

For $M_n = M_p$ the latter reduces to

$$p_f = (0, \boldsymbol{k})\,. \tag{3.1.19}$$

At the end of this section we shall consider the inverse reaction, radiative n-p capture ($n + p \to d + \gamma$), since it is related to deuteron photodisintegration via the principle of detailed balance. Here the variables in the lab frame are

$$p_p^L = (M_p, \boldsymbol{0}) \tag{3.1.20}$$

for the proton at rest,

$$p_n^L = (E_n^L, -\boldsymbol{k}^L) \tag{3.1.21}$$

for the incoming neutron, and in the cm system

$$p_{p/n} = (E_{p/n}, \pm\boldsymbol{k}) \tag{3.1.22}$$

$$p_\gamma = (\omega, \boldsymbol{\omega}) \tag{3.1.23}$$

$$p_d = (E_d, -\boldsymbol{\omega})\,, \tag{3.1.24}$$

where $\boldsymbol{k}$ is given by (3.1.11) using

$$W = (M_n^2 + M_p^2 + 2M_pE_n^L)^{1/2}\,, \tag{3.1.25}$$

since the proton is at rest in the lab frame. Comparing (3.1.25) with (3.1.10) one finds the relation between the neutron lab kinetic energy in $n + p \to d + \gamma$ and the photon lab energy in $\gamma + d \to n + p$ corresponding to the same total cm energy W

$$\omega^L = \frac{M_p}{M_d}\,E_{n,kin}^L + \frac{1}{2M_d}((M_n + M_p)^2 - M_d^2)\,. \tag{3.1.26}$$

3.2 The T-matrix

The general expression for the differential cross section of a two-body reaction ($p_1^i + p_2^i \rightarrow p_1^f + p_2^f$) is given by

$$d\sigma = \frac{(2\pi)^{10}}{4p_{10}^i p_{20}^i v_{12}} \sum_f \frac{d^3p_1^f}{2p_{10}^f} \frac{d^3p_2^f}{2p_{20}^f} \delta^{(4)}(P_f - P_i) \, |T_{fi}|^2 \,, \tag{3.2.1}$$

where the T-matrix is related to the scattering matrix by

$$S_{fi} = \langle f|i\rangle + i(2\pi)^4 \delta^{(4)}(P_f - P_i) T_{fi} \,. \tag{3.2.2}$$

The relative velocity in the incoming channel is denoted by v_{12}. Covariant normalization of the states is used

$$\langle \boldsymbol{p}|\boldsymbol{p}'\rangle = 2p_0 \delta^{(3)}(\boldsymbol{p} - \boldsymbol{p}') \,. \tag{3.2.3}$$

Evaluating (3.2.1) in the cm frame for deuteron photodisintegration, one obtains for the differential cross section

$$\frac{d\sigma}{d\Omega} = \frac{(2\pi)^{10}}{(4W)^2} \frac{k}{\omega} |T_{fi}|^2 \,, \tag{3.2.4}$$

where the T-matrix is given in first-order perturbation theory by

$$T_{fi} = \sqrt{\frac{\alpha}{2\pi^2}} \langle \boldsymbol{p}_f, W | J_\lambda(0) | p_d \rangle \tag{3.2.5}$$

with

$$J_\lambda(0) = \varepsilon^\mu(\lambda) J_\mu(0) \,. \tag{3.2.6}$$

The elementary charge has been separated from the current density operator $J_\mu(x)$. The photon polarization vector is denoted by $\varepsilon^\mu(\lambda)$ and $\alpha = e^2/4\pi$ is the fine-structure constant. The final p-n state $|\boldsymbol{p}_f, W\rangle$ has total three-momentum zero and is characterized by the asymptotic relative momentum $\boldsymbol{p}_f$ as given in (3.1.18). The initial deuteron state has total momentum $p_d = (E_d, -\boldsymbol{\omega})$.

For later convenience we will define a new T-matrix T^γ incorporating the kinematical factors of (3.2.4) by setting

$$\frac{d\sigma}{d\Omega} = |T_{fi}^\gamma|^2 \tag{3.2.7}$$

with

$$T_{fi}^\gamma = \frac{(2\pi)^4}{4W} \sqrt{\frac{2\alpha k}{\omega}} \langle \boldsymbol{p}_f, W | J_\lambda(0) | p_d \rangle \,. \tag{3.2.8}$$

Eq. (3.2.7) is valid for definite polarization and spin projection.

It is useful for the evaluation of the current matrix element in (3.2.8) to separate the center-of-mass motion and to describe the internal wave functions of the initial and final states in terms of the rest frame one by introducing a boost operator $U(\boldsymbol{P})$ which acts on the intrinsic rest frame wave function

$$\psi_{d\boldsymbol{P}} = U(\boldsymbol{P})\psi_d \tag{3.2.9}$$

where $\boldsymbol{P}$ denotes the cm momentum of the two-body system. In the non-relativistic limit $U(\boldsymbol{P})$ is the unit operator since the internal wave function does not depend on the cm motion. This is not any more the case if relativistic effects are considered as will be discussed in section 4.3.

Separating out the center-of-mass motion by integrating over the cm coordinates gives then

$$\langle P_f|J_\lambda(0)|P_d\rangle = \frac{1}{(2\pi)^3}\,\langle f|\tilde{J}_\lambda(\omega,\boldsymbol{Q})|d\rangle^{\mathrm{cm}}\,, \tag{3.2.10}$$

where $|d\rangle^{\mathrm{cm}}$ and $|f\rangle^{\mathrm{cm}}$ refer to intrinsic rest-frame wave functions, and

$$\tilde{J}_\lambda(\omega,\boldsymbol{Q}) = \int d^3R\, U^\dagger(\boldsymbol{P}_f)J_\lambda(0)U(\boldsymbol{P}_d)e^{-i\omega\cdot\boldsymbol{R}} \tag{3.2.11}$$

is the Fourier transform of the current operator modified by the boost operators. It is also a function of $\boldsymbol{Q}=\boldsymbol{P}_f+\boldsymbol{P}_d$ and depends on intrinsic, i.e. rest-frame variables only.

Until now, we have not specified the spin quantum numbers of the initial deuteron and the final p-n states. In the non-relativistic limit one chooses an arbitrary quantization axis. Then the initial deuteron state is defined by its spin projection m_d with respect to this axis. The final p-n state may be characterized by its total spin s (=0 or 1) and its projection m_s on the relative momentum $\boldsymbol{k}$. Furthermore, choosing non-covariant normalization one obtains then in transverse gauge ($J_\lambda=-\boldsymbol{\varepsilon}(\lambda)\cdot\boldsymbol{J}$)

$$T^\gamma_{sm_s\lambda m_d(\hat{\boldsymbol{n}})} = \frac{2\pi}{W}(\alpha kE_pE_nE_d/\omega)^{1/2}\,\langle\boldsymbol{k}sm_s|\tilde{J}_\lambda(\omega,\boldsymbol{Q})|m_d\rangle^{\mathrm{cm}}_{n.c.}\,, \tag{3.2.12}$$

where $m_d(\hat{\boldsymbol{n}})$ indicates that the deuteron spin projection refers to the quantization axis $\hat{\boldsymbol{n}}$ and $\tilde{\boldsymbol{J}}(\omega,\boldsymbol{Q})$ is the Fourier transform of the current density referring to intrinsic variables only.

We would like to point out that in q-congruent systems, i.e., $\boldsymbol{Q}$ proportional to $\boldsymbol{q}$ like, e.g., in the overall cm system ($\boldsymbol{P}_f=0$) the cm current

$$\boldsymbol{J}_{\mathrm{cm}}(\boldsymbol{q},\boldsymbol{Q}) = \frac{\boldsymbol{Q}}{4M}\rho(\boldsymbol{q}) = \frac{\boldsymbol{q}}{4M}\rho(\boldsymbol{q}) \tag{3.2.13}$$

is purely longitudinal and thus drops out. This observation is important with respect to some remarks on the role of the cm motion in (JaW 84a, MoR 87) which we shall discuss in section 4.1. Therefore, in the non-relativistic limit only the intrinsic current density contributes in (3.2.12), which depends on ω only.

The wave functions $|m_d\rangle^{\mathrm{cm}}_{n.c.}$ and $|\boldsymbol{k}sm_s\rangle^{\mathrm{cm}}_{n.c.}$ describe the rest-frame intrinsic motion of deuteron and p-n states, respectively, in non-covariant normalization. For $M_n=M_p$ one has with $E_n=E_p=W/2$

$$T^\gamma_{sm_s\lambda m_d} = \pi(\alpha kE_d/\omega)^{1/2}\langle\boldsymbol{k}sm_s|J_\lambda(\omega)|m_d\rangle^{\mathrm{cm}}_{n.c.}\,, \tag{3.2.14}$$

which corresponds to the T-matrix of de Swart (DSw 59) and Partovi (Par 64), if one approximates E_d by $2M$ and chooses the quantization axis for the deuteron spin projections parallel to $\boldsymbol{k}$. This choice is useful for the numerical evaluation. However, it is not convenient if deuteron orientation is considered. In this case a better choice is to take the direction of the incoming photon as quantization axis. Then one has to transform the T-matrix according to

$$T^\gamma_{sm_s\lambda m_d(\hat{\omega})} = \sum_{m'_d} T^\gamma_{sm_s\lambda m'_d(\hat{\boldsymbol{k}})}\, D^1_{m'_dm_d}(0,-\Theta,-\Phi)\,, \tag{3.2.15}$$

where we use the convention of Rose (Ros 57) for the rotation matrix $D^j_{m'm}$. The spherical angles Θ and Φ define the direction of the relative p-n momentum $\boldsymbol{k}$ in the cm system with respect to a frame of reference associated with the incoming photon having its z-axis in the direction of the photon momentum $\boldsymbol{\omega}$ (see fig. 3.2.1). Thus the rotation, which carries $\boldsymbol{k}$ into $\boldsymbol{\omega}$ is given by the Euler angles $(\alpha, \beta, \gamma) = (\Phi, \Theta, 0)$. In fig. 3.2.1 we have also specified a frame of reference (x', y', z') with z'-axis along $\boldsymbol{k}$ in which the matrix elements will be evaluated. It serves as frame of reference for the nucleon polarization components and is chosen according to the Madison convention (Mad 71) for the outgoing proton but not for the outgoing neutron.

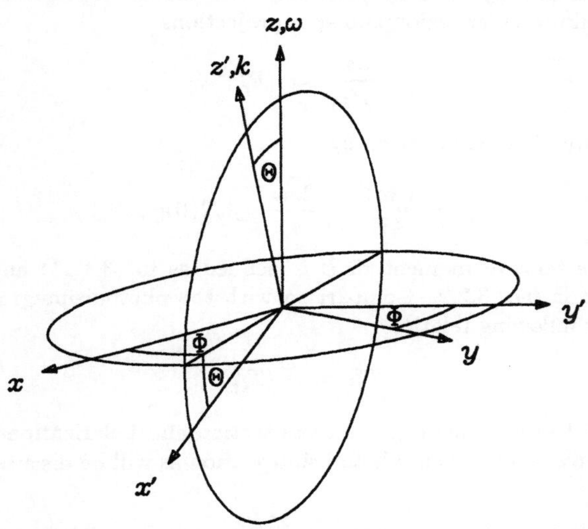

Fig. 3.2.1: Definition of reference frames for deuteron photodisintegration in the cm system. Incoming-photon momentum $\boldsymbol{\omega}$ along z-axis and relative p-n momentum $\boldsymbol{k}$ along z'-axis, characterized by Θ and Φ.

The T-matrix has the following form (see (3.5.1))

$$T^{\gamma}_{sm_s\lambda m_d}(\Theta, \Phi) = e^{i(\lambda+m_d)\Phi} t_{sm_s\lambda m_d}(\Theta) . \tag{3.2.16}$$

As indicated, the reduced t-matrix $t_{sm_s\lambda m_d}$ depends on Θ only. Thus, the azimuthal Φ-dependence is separated and explicitly given. Parity conservation leads to the following symmetry relation

$$t_{s-m_s-\lambda-m_d} = (-)^{1+s+m_s+\lambda+m_d} t_{sm_s\lambda m_d} . \tag{3.2.17}$$

For a relativistic treatment the helicity formalism is often used. Denoting the helicities of deuteron, proton and neutron by λ_d, λ_p and λ_n, respectively, one obtains for the T-matrix

$$T^{\gamma}_{\lambda_p\lambda_n\lambda\lambda_d} = \frac{(2\pi)^4}{W}\sqrt{\frac{2\alpha k}{\omega}}\langle \boldsymbol{k}, \lambda_p\lambda_n | J_\lambda(0) | -\boldsymbol{\omega}, \lambda_d\rangle . \tag{3.2.18}$$

Choosing as above the quantization axis for the deuteron states parallel to $\boldsymbol{\omega}$, one has the following relation to the T-matrix in the helicity basis

$$T^{\gamma}_{\lambda_p\lambda_n\lambda\lambda_d} = e^{-i(\lambda_p-\lambda_n)\Phi} \sum_{sm_s m_d} (\tfrac{1}{2}\lambda_p \tfrac{1}{2} - \lambda_n | sm_s) T^{\gamma}_{sm_s\lambda-\lambda_d} \,. \tag{3.2.19}$$

Counting all possible combinations of initial- and final-state polarization one finds 24 complex amplitudes of which only 12 are independent if parity is conserved according to (3.2.17). In this case, one needs 23 observables for a complete determination of these amplitudes, since one phase remains undetermined.

Before discussing the general expressions for cross section and polarization observables, we will shortly consider again the inverse process, radiative n-p capture $n+p \to d+\gamma$. In analogy to (3.2.7), one may write for the capture cross section in the cm frame for definite polarization and spin projections

$$\frac{d\sigma^c}{d\Omega_\gamma} = |T^c_{fi}|^2 \,, \tag{3.2.20}$$

where the capture T-matrix is given by

$$T^c_{fi} = \frac{(2\pi)^4}{4W} \sqrt{\frac{2\alpha\omega}{k}} \langle p_d | J^*_\lambda(0) | p_i, W \rangle \,. \tag{3.2.21}$$

The initial state relative momentum $\boldsymbol{k}$ is defined as in (3.1.11) and the coordinate system is shown in fig. 3.2.2. Comparison with the photodisintegration T-matrix in (3.2.8) gives the following relation

$$T^c_{fi} = \frac{\omega}{k} (T^{\gamma}_{if})^* \,, \tag{3.2.22}$$

which will allow to relate the capture cross section and polarization quantities to the corresponding inverse process in photodisintegration as will be discussed in section 3.4.

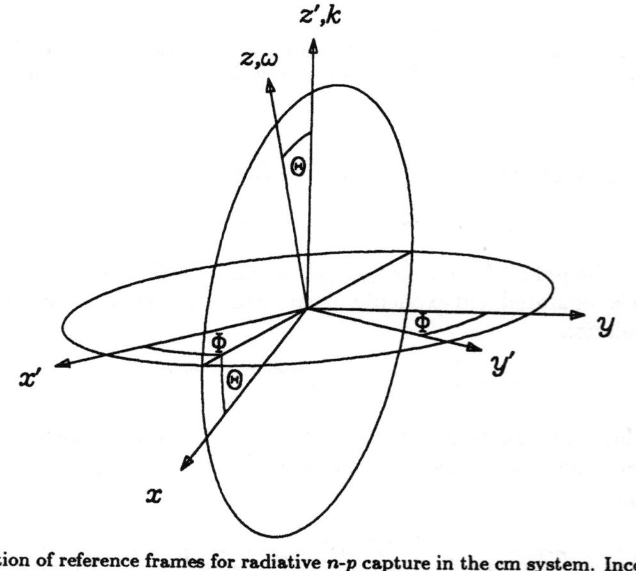

Fig. 3.2.2: Definition of reference frames for radiative n-p capture in the cm system. Incoming relative p-n momentum $\boldsymbol{k}$ along z'-axis and outgoing photon momentum $\boldsymbol{\omega}$ along z-axis, characterized by Θ and Φ.

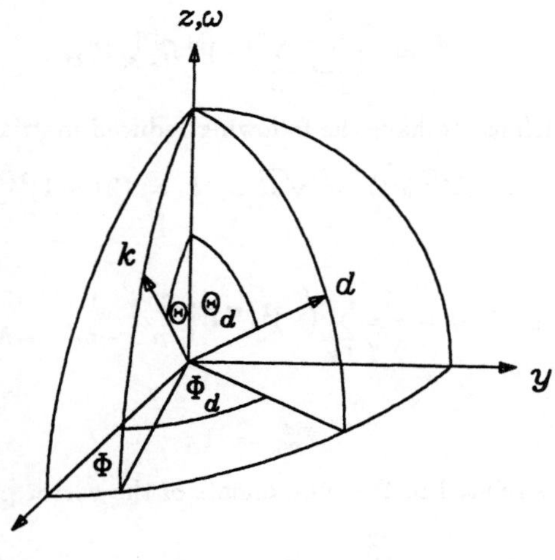

Fig. 3.2.3: Definition of the deuteron orientation axis characterized by the angles of Θ_d and Φ_d.

3.3 Density matrices

In order to derive the general expressions for cross section and polarization quantities, we shall briefly review the density matrices of photon, nucleon and deuteron (Rob 74).

The nucleon density matrix is a 2×2 matrix and has the general form

$$\rho^N = \frac{1}{2}(1 + \boldsymbol{P}^N \cdot \boldsymbol{\sigma}) , \tag{3.3.1}$$

where $|\boldsymbol{P}^N|$ gives the total degree of polarization and P_i^N the polarization component of the nucleon with respect to the i-th axis. As mentioned in section 3.2 the reference frame is specified in fig. 3.2.1 and agrees with the Madison convention for the proton polarization. However, one should keep in mind that for the neutron polarization the x- and y-components have the opposite sign with respect to the Madison convention.

The photon density has formally the same structure

$$\rho^\gamma = \frac{1}{2}(1 + \boldsymbol{P}^\gamma \cdot \boldsymbol{\sigma}) , \tag{3.3.2}$$

where again $|\boldsymbol{P}^\gamma|$ characterizes the total degree of polarization. However, the meaning of the components of $\boldsymbol{P}^\gamma$ is different. Taking the z-axis along the photon momentum, then $P_c^\gamma = P_z^\gamma$ is the difference of right to left circularly polarized photons, i.e., $|P_c^\gamma|$ describes the degree of circular polarization being right or left according to whether $P_c^\gamma > 0$ or < 0, respectively. The degree of linear polarization is given by $P_l^\gamma = ((P_x^\gamma)^2 + (P_y^\gamma)^2)^{1/2}$. By a rotation around the z-axis, one can always reach $P_y^\gamma = 0$. Then, one has linear polarization along the x-axis for $P_x^\gamma < 0$ and along the y-axis for $P_x^\gamma > 0$. From now on we shall fix the coordinate system associated with the incoming photon in fig. 3.2.1 by demanding maximal linear polarization along the x-axis, i.e., $P_x^\gamma < 0$.

Since the deuteron has spin one, its density matrix can be decomposed into scalar, vector and tensor operators ($\Omega_M^{[I]}$, $I = 0, 1, 2$) with corresponding orientation parameters P_{IM}^d

$$\rho^d = \frac{1}{3}\sum_{I=0}^{2}\sum_{M}(-)^M \Omega^{[I]}_{-M} P^d_{IM} \,. \tag{3.3.3}$$

The operators are defined to have the following reduced matrix elements

$$\langle 1 \parallel \Omega^{[I]} \parallel 1\rangle = \sqrt{3}\hat{I}\,, \quad \hat{I} = (2I+1)^{1/2}\,. \tag{3.3.4}$$

Thus, one has

$$\rho^d_{m'm} = \langle 1m'|\rho^d|1m\rangle = \frac{1}{\sqrt{3}}\sum_{IM}(-)^{1-m}\hat{I}\begin{pmatrix} 1 & 1 & I \\ m & -m' & -M \end{pmatrix} P^d_{IM}\,, \tag{3.3.5}$$

where

$$P^d_{00} = 1\,. \tag{3.3.6}$$

Furthermore, P^d_{1M} is related to the components of the vector polarization

$$P^d_{1M} = \sqrt{\frac{3}{2}}(\delta_{M0}P_z - M(P_x + iMP_y)/\sqrt{2}) \tag{3.3.7}$$

and P^d_{2M} to the various tensor polarizations of the deuteron

$$P^d_{20} = \frac{1}{\sqrt{2}}P_{zz} \tag{3.3.8}$$

$$P^d_{2\pm1} = \mp\frac{1}{\sqrt{3}}(P_{xz} \pm iP_{yz}) \tag{3.3.9}$$

$$P^d_{2\pm2} = \frac{1}{2\sqrt{3}}(P_{xx} - P_{yy}) \pm \frac{2i}{\sqrt{3}}P_{xy}\,. \tag{3.3.10}$$

For presently available sources of oriented deuterons (Mey 85a,b) the density matrix is diagonal with respect to a certain orientation axis $\hat{d}$ characterized by the angles Θ_d and Φ_d with respect to the chosen frame of reference (see fig. 3.2.3), i.e.,

$$\rho^d_{m'm} = p_m\delta_{m'm}\,, \tag{3.3.11}$$

where p_m is the probability of finding a deuteron with spin projection m on $\hat{d}$. Then the only non-vanishing orientation parameters with respect to $\hat{d}$ are

$$P^d_1 = P^d_{10} = \sqrt{\frac{3}{2}}(p_1 - p_{-1})\,, \tag{3.3.12}$$

$$P^d_2 = P^d_{20} = \frac{1}{\sqrt{2}}(p_1 + p_{-1} - 2p_0) = \frac{1}{\sqrt{2}}(1 - 3p_0)\,. \tag{3.3.13}$$

The orientation parameters with respect to the photon momentum as quantization axis are obtained by a rotation

$$P^d_{IM}(\omega) = P^d_I D^I_{0M}(0, -\Theta_d, -\Phi_d) = P^d_I e^{iM\Phi} d^I_{M0}(\Theta_d)\,. \tag{3.3.14}$$

The total initial-state density matrix is then given by the direct product of photon and deuteron density matrices $\rho_i = \rho^\gamma \otimes \rho^d$. For the inverse process of n-p capture the initial-state density matrix is obtained from the direct product of proton and neutron density matrices $\rho^c_i = \rho^p \otimes \rho^n$.

3.4 Cross section and polarization observables

The differential cross section for arbitrary photon and deuteron polarization is given by

$$\begin{aligned}\frac{d\sigma^\gamma(\rho_i)}{d\Omega} &= Tr(T^\gamma \rho_i T^{\gamma\dagger}) \\ &= \sum_{sm_s\lambda\lambda' m_d m_d'} T^\gamma_{sm_s\lambda m_d}\rho^\gamma_{\lambda\lambda'}\rho^d_{m_d m_d'} T^{\gamma^*}_{sm_s\lambda' m_d'}\end{aligned} \tag{3.4.1}$$

and the polarization $P_\alpha(j)$ of one nucleon ($j = 1$ or 2) of the final state by

$$\begin{aligned}P_\alpha(j)\frac{d\sigma^\gamma}{d\Omega} &= Tr(T^\gamma \rho_i T^{\gamma\dagger}\sigma_\alpha(j)) \\ &= \sum_{sm_s s'm_s'\lambda\lambda' m_d m_d'} T^\gamma_{sm_s\lambda m_d}\rho^\gamma_{\lambda\lambda'}\rho^d_{m_d m_d'} T^{\gamma^*}_{s'm_s'\lambda' m_d'} \\ &\qquad \langle s'm_s'|\sigma_\alpha(j)|sm_s\rangle \,.\end{aligned} \tag{3.4.2}$$

If one observes the polarization of both nucleons in the final state, then one has an analogous expression replacing $P_\alpha(j)$ by $P_{\alpha_1\alpha_2}(1,2)$ and $\sigma_\alpha(j)$ by $\sigma_{\alpha_1}(1)\ \sigma_{\alpha_2}(2)$. Similar expressions hold in the helicity basis replacing the sums over ($s\ m_s\ m_d$) by sums over the helicities ($\lambda_p\ \lambda_n\ \lambda_d$) and $\rho_{m_d m_d'}$, by $\rho_{-\lambda_d-\lambda_d'}$.

Separating out the photon and deuteron polarization parameters, one can write

$$\frac{d\sigma^\gamma(\rho_i)}{d\Omega} = \frac{1}{2}\sum_{\lambda,IM}(A^\gamma_{\lambda\lambda IM}(1) - P^\gamma_l A^\gamma_{\lambda-\lambda IM}(1) + P^\gamma_c \lambda A^\gamma_{\lambda\lambda IM}(1))P^{d^*}_{IM}\,. \tag{3.4.3}$$

For the photon density matrix we have chosen the special coordinate system for which $P^\gamma_y = 0$ and have set $P^\gamma_x = -P^\gamma_l$, where P^γ_l denotes the degree of linearly polarized photons (see section 3.3). Here we have introduced for an arbitrary spin operator $\hat\Omega$

$$\begin{aligned}A^\gamma_{\lambda\lambda' IM}(\hat\Omega) &= \frac{\hat I}{\sqrt3}\sum_{sm_s s'm_s' m_d m_d'}(-)^{1-m_d'}\begin{pmatrix}1 & 1 & I\\ m_d' & -m_d & M\end{pmatrix} \\ &\qquad T^\gamma_{sm_s\lambda m_d}T^{\gamma^*}_{s'm_s'\lambda' m_d'}\ \langle s'm_s'|\hat\Omega|sm_s\rangle\end{aligned} \tag{3.4.4}$$

or in the helicity basis

$$\begin{aligned}A^\gamma_{\lambda\lambda' IM}(\hat\Omega) &= \frac{\hat I}{\sqrt3}\sum_{\lambda_p\lambda_n\lambda_d\lambda_p'\lambda_n'\lambda_d'}(-)^{1-\lambda_d'}\begin{pmatrix}1 & 1 & I\\ -\lambda_d' & \lambda_d & M\end{pmatrix}T^\gamma_{\lambda_p\lambda_n\lambda\lambda_d}T^{\gamma^*}_{\lambda_p'\lambda_n'\lambda'\lambda_d'} \\ &\qquad \langle \tfrac12\lambda_p'\tfrac12\lambda_n'|\hat\Omega|\tfrac12\lambda_p\tfrac12 - \lambda_n\rangle\,.\end{aligned} \tag{3.4.5}$$

In a similar manner one can decompose the polarization of the j-th nucleon writing

$$P_\alpha(j)\frac{d\sigma}{d\Omega} = \frac{1}{2}\sum_{\lambda IM}(A^\gamma_{\lambda\lambda IM}(\sigma_\alpha(j)) - P^\gamma_l A^\gamma_{\lambda-\lambda IM}(\sigma_\alpha(j)) + P^\gamma_c\lambda A^\gamma_{\lambda\lambda IM}(\sigma_\alpha(j)))P^{d^*}_{IM}\,. \tag{3.4.6}$$

For the observation of the polarization of both nucleons, one obtains a corresponding expression by replacing in (3.4.6) the operator $\sigma_\alpha(j)$ by $\sigma_{\alpha_1}(1)\sigma_{\alpha_2}(2)$.

On the basis of the explicit Φ-dependence of the T-matrix in (3.2.16) and the symmetry relation (3.2.17) one can derive more compact expressions. As is shown in Are 88a all observables can be expressed in terms of the quantities

$$v/w_{IM}^{s'sS\sigma} = \hat{I} \sum_{m'_s m_s m'_d m_d} (-)^{1-m_d+s'-m'_s} \begin{pmatrix} 1 & 1 & I \\ m'_d & -m_d & M \end{pmatrix} \begin{pmatrix} s' & s & S \\ m'_s & -m_s & -\sigma \end{pmatrix} t^*_{s'm'_s 1 m'_d} t_{sm_s \pm 1 m_d} \,, \tag{3.4.7}$$

which have the symmetry properties as obtained from (3.2.17)

$$(v_{IM}^{s'sS\sigma})^* = (-)^{s+s'+\sigma+M} v_{IM}^{ss'S-\sigma} \tag{3.4.8}$$

$$(w_{IM}^{s'sS\sigma})^* = (-)^{s+s'+S+I} w_{IM}^{ss'S\sigma} \,. \tag{3.4.9}$$

For the differential cross section in the cm frame one obtains

$$\begin{aligned} \frac{d\sigma}{d\Omega} = \frac{d\sigma_0}{d\Omega}[1 &+ P_l^\gamma \Sigma^l(\Theta)\cos 2\Phi \\ &+ \sum_{I=1,2} P_I^d\{\sum_{M\geq 0}(T_{IM}(\Theta)\cos(M(\Phi_d-\Phi)-\delta_{I1}\pi/2) \\ &+ P_c^\gamma T_{IM}^c(\Theta)\sin(M(\Phi_d-\Phi)+\delta_{I1}\pi/2))d_{M0}^I(\Theta_d) \\ &+ P_l^\gamma \sum_{M=-I}^{I} T_{IM}^l(\Theta)\cos(\Psi_M-\delta_{I1}\pi/2)d_{M0}^I(\Theta_d)\}] \,, \end{aligned} \tag{3.4.10}$$

where

$$\Psi_M = M(\Phi_d-\Phi)+2\Phi \,. \tag{3.4.11}$$

Here the unpolarized cross section and the various photon and target asymmetries are given by

$$\frac{d\sigma_0}{d\Omega} = V_{00} \tag{3.4.12}$$

$$\Sigma^l(\Theta)\frac{d\sigma_0}{d\Omega} = -W_{00} \,, \tag{3.4.13}$$

$$T_{11}(\Theta)\frac{d\sigma_0}{d\Omega} = 2\ \mathrm{Im}\ V_{11} \,, \tag{3.4.14}$$

$$T_{2M}(\Theta)\frac{d\sigma_0}{d\Omega} = (2-\delta_{M0})\ \mathrm{Re}\ V_{2M}, \qquad M\geq 0 \tag{3.4.15}$$

$$T_{1M}^C(\Theta)\frac{d\sigma_0}{d\Omega} = (2-\delta_{M0})\ \mathrm{Re}\ V_{1M} \,, \qquad M\geq 0 \tag{3.4.16}$$

$$T_{2M}^C(\Theta)\frac{d\sigma_0}{d\Omega} = 2\ \mathrm{Im}\ V_{2M} \,, \qquad M\geq 0 \tag{3.4.17}$$

$$T_{1M}^l(\Theta)\frac{d\sigma_0}{d\Omega} = iW_{1M} \,, \tag{3.4.18}$$

$$T_{2M}^l(\Theta)\frac{d\sigma_0}{d\Omega} = -W_{2M} \,, \tag{3.4.19}$$

where V_{IM} and W_{IM} are defined by

$$V/W_{IM} = \frac{1}{\sqrt{3}}\sum_s \hat{s} v/w_{IM}^{ss00} \,. \tag{3.4.20}$$

For details see Are 88a. Note that the photon asymmetry with respect to circularly polarized photons vanishes for unoriented deuterons because of the rotational symmetry of the initial state. For the nucleon polarization observables one obtains similar expressions which have been derived in Are 88a and are listed in appendix A.

These expressions for differential cross section and one- and two-nucleon polarization contain the complete information on the photodisintegration process. But the resulting set of 288 observables is certainly overcomplete, since a set of 23 independent observables is already sufficient for the determination of the T-matrix as we have mentioned before. A first reduction is achieved by inverting (3.4.7) which allows to express the products of t-matrix elements in terms of either v's or w's. Since these in turn can be expressed by 144 observables for unpolarized and circularly polarized photons or by 144 observables for linearly polarized photons, respectively, one obtains linear relations between these two sets of 144 observables. Furthermore, fixing one t-matrix element one can derive explicit sets of 23 observables, in terms of which the t-matrix elements can be expressed (ArS 90).

For the total disintegration cross section one obtains

$$\begin{aligned} \sigma_{tot}(P^\gamma, P_d) &= \sigma^0_{tot}(1 + P^d_2 \tau^0_{20} P_2(\cos\Theta_d) \\ &\quad + P^\gamma_c P^d_1 \tau^c_{10} \cos\Theta_d \\ &\quad + P^\gamma_l P^d_2 \tau^l_{22} d^2_{20}(\Theta_d) \cos 2\Phi_d)\,, \end{aligned} \tag{3.4.21}$$

where

$$\sigma^0_{tot} = 2\pi \int d(\cos\Theta) \frac{d\sigma_0}{d\Omega} \tag{3.4.22}$$

$$\tau^\alpha_{IM}\sigma^0_{tot} = 2\pi \int d(\cos\Theta) T^\alpha_{IM} \frac{d\sigma_0}{d\Omega} \tag{3.4.23}$$

with $\alpha = 0, c, l$.

One readily sees that for unpolarized photons there is only a contribution from the deuteron tensor polarization P^d_2. As one would expect, dependence on circular photon polarization exists only for vector polarized deuterons not perpendicular to the incoming-photon direction while dependence on linear polarization exists only for non-vanishing deuteron tensor polarization P^d_2.

At the end of this section we shall consider n-p capture. The cross section including beam and target polarization is given by

$$\begin{aligned} \frac{d\sigma^c}{d\Omega_\gamma} &= Tr(T^c \rho^p \rho^n T^{c\dagger}) \\ &= \frac{1}{4} \sum_{sm_s s'm'_s \lambda m_d} T^c_{\lambda m_d s m_s} T^{c*}_{\lambda m_d s' m'_s} \\ &\qquad \langle s m_s | (1 + \boldsymbol{p}_p \cdot \boldsymbol{\sigma}(1))(1 + \boldsymbol{p}_n \cdot \boldsymbol{\sigma}(2)) | s' m'_s \rangle \\ &= \frac{d\sigma^c_0}{d\Omega_\gamma}(1 + \boldsymbol{p}_p \cdot \boldsymbol{A}(p) + \boldsymbol{p}_n \cdot \boldsymbol{A}(n) \\ &\qquad + \sum_{k,l} p_{p,k} p_{n,l} A_{kl}(pn))\,, \end{aligned} \tag{3.4.24}$$

which defines the various target and beam asymmetries for a polarized beam and/or target. The beam and target polarizations are denoted by $\boldsymbol{p}_n$ and $\boldsymbol{p}_p$, respectively.

The capture cross section for an unpolarized beam and target is denoted by $d\sigma_0^c/d\Omega_\gamma$. Using the relation (3.2.22) between the capture and the disintegration T-matrix one can relate the capture cross section and asymmetries to the disintegration cross section and nucleon polarization observables

$$\frac{d\sigma_0^c}{d\Omega_\gamma} = \frac{3}{2}\left(\frac{\omega}{k}\right)^2 \frac{d\sigma_0}{d\Omega} \tag{3.4.25}$$

$$A(p/n) = P^{0,00}(p/n) \tag{3.4.26}$$

$$A_{x_i x_j}(pn) = P^{0,00}_{x_i x_j} , \tag{3.4.27}$$

where in (3.4.25) the factor 3/2 arises from the different statistical weights of the initial states. Note that here the same convention for the coordinate systems for the nucleon polarization in n-p capture as in $d(\gamma, N)N$ is used. Thus, one can determine the outgoing nucleon polarization in $d(\gamma, N)N$ from a corresponding capture experiment with polarized beam and/or target.

3.5 Multipole decomposition and angular distributions

In order to exhibit explicitly the angular dependence of the differential cross section and polarization observables, it is convenient to make a partial-wave expansion or multipole decomposition of the transition amplitude

$$T^\gamma_{sm_s\lambda m_d} = \sum_{Lljm_j\mu} \frac{\hat{l}}{\hat{j}}(1m_d L\lambda|jm_j)(l0sm_s|jm_s)\mathcal{O}^{L\lambda}(\mu jls)D^j_{m_s m_j}(R) , \tag{3.5.1}$$

where R denotes the rotation of the relative p-n momentum $\boldsymbol{k}$ into the photon momentum $\boldsymbol{\omega}$ by the Euler angles $(0, -\Theta, -\Phi)$ with Θ and Φ being the spherical angles of $\boldsymbol{k}$ with respect to the reference system $\boldsymbol{\omega}$. The quantity denoted by the symbol $\mathcal{O}^{L\lambda}(\mu jls)$ contains the reduced electric and magnetic multipole transition moments of order L between the deuteron and the final-state partial wave $|\mu jm_j\rangle$. It will be specified later (see section 4.1). From (3.5.1) follows directly helicity conservation for $\Theta = 0$ and π, i.e., the T-matrix vanishes at $\Theta = 0$ and π unless $\lambda + m_d = \pm m_s$, respectively.

It is obvious from (3.5.1) that the Φ-dependence can be separated out, a fact already used in (3.2.16). The resulting reduced t-matrix is then inserted into (3.4.7). With the help of angular momentum sum rules considerable simplification is achieved. In detail one finds for (3.4.20)

$$V/W_{IM} = \sum_K V/W^K_{IM} d^K_{M-(1\mp1),0}(\Theta), \tag{3.5.2}$$

where

$$\begin{aligned} V/W^K_{IM} = {} & \frac{(-)^{1+M}}{\sqrt{3}}\hat{I}\hat{K}^2 \sum_J \hat{J}^2 \begin{pmatrix} J & I & K \\ -1\pm1 & M & 1\mp1-M \end{pmatrix} \\ & \sum_{L\mu jlsL'\mu'j'l'} (-)^{L+j'+s}\hat{l}\hat{l}'\hat{j}\hat{j}' \begin{pmatrix} l & l' & K \\ 0 & 0 & 0 \end{pmatrix} \begin{pmatrix} L & L' & J \\ \pm1 & -1 & 1\mp1 \end{pmatrix} \\ & \begin{Bmatrix} l' & l & K \\ j & j' & s \end{Bmatrix} \begin{Bmatrix} j & j' & K \\ 1 & 1 & I \\ L & L' & J \end{Bmatrix} \mathcal{O}^{L'1^*}(\mu' j' l' s)\mathcal{O}^{L\pm1}(\mu jls) . \end{aligned} \tag{3.5.3}$$

In particular, for $I = 0$ (unpolarized deuteron) one has

$$\frac{d\sigma}{d\Omega} = \sum_K V_{00}^K P_K(\cos\Theta) , \tag{3.5.4}$$

where

$$V_{00}^K = \frac{\hat{K}^2}{3} \sum_{L\mu jlsL'\mu' j'l'} (-)^{L'+L+j'+j+s}\hat{l}\hat{l}'\hat{j}\hat{j}' \begin{pmatrix} l & l' & K \\ 0 & 0 & 0 \end{pmatrix} \begin{pmatrix} L & L' & K \\ 1 & -1 & 0 \end{pmatrix}$$
$$\begin{Bmatrix} l' & l & K \\ j & j' & s \end{Bmatrix} \begin{Bmatrix} L' & L & K \\ j & j' & 1 \end{Bmatrix} \mathcal{O}^{L'1*}(\mu' j' l' s)\mathcal{O}^{L1}(\mu j l s) , \tag{3.5.5}$$

which has already been given by Kawaguchi (Kaw 58) and later rederived by Cambi et al. (CaM 82a). It has been emphasized in CaM 82a that the expansion (3.5.4) of the differential cross section in terms of Legendre polynomials is much more appropriate for fitting experimental data than the expansion of de Swart (DSw 59) and Partovi (Par 64) in terms of $(a_n + b_n \cos\Theta)\sin^{2n}\Theta$, which has often been used in the past. The reason for this is the fact that the coefficients of an expansion in orthogonal functions are uniquely determined whereas the coefficients of the de Swart-Partovi expansion are not free of ambiguities. Relations between the coefficients of the two expansions have been derived in CaM 82a and can be found in the appendix B.

Then, one obtains for the various contributions to the differential cross section of (3.4.10)

$$\Sigma^l(\Theta)\frac{d\sigma_0}{d\Omega} = -\sum_K W_{00}^K d_{20}^K(\Theta) , \tag{3.5.6}$$

$$T_{11}(\Theta)\frac{d\sigma_0}{d\Omega} = 2\sum_K \text{Im } V_{11}^K d_{10}^K(\Theta) , \tag{3.5.7}$$

$$T_{2M}(\Theta)\frac{d\sigma_0}{d\Omega} = (2-\delta_{MO})\sum_K \text{Re } V_{2M}^K d_{MO}^K(\Theta), \quad M \geq 0 , \tag{3.5.8}$$

$$T_{1M}^c(\Theta)\frac{d\sigma_0}{d\Omega} = (2-\delta_{MO})\sum_K \text{Re } V_{1M}^K d_{MO}^K(\Theta), \quad M \geq 0 , \tag{3.5.9}$$

$$T_{2M}^c(\Theta)\frac{d\sigma_0}{d\Omega} = 2\sum_K \text{Im } V_{2M}^K d_{MO}^K(\Theta), \quad M \geq 0 , \tag{3.5.10}$$

$$T_{1M}^l(\Theta)\frac{d\sigma_0}{d\Omega} = i\sum_K W_{1M}^K d_{M-2,0}^K(\Theta) , \tag{3.5.11}$$

$$T_{2M}^l(\Theta)\frac{d\sigma_0}{d\Omega} = -\sum_K W_{2M}^K d_{M-2,0}^K(\Theta) , \tag{3.5.12}$$

where the d-functions of the rotation matrices can be expressed in terms of Legendre functions

$$d_{MO}^K(\Theta) = \sqrt{\frac{(K-M)!}{(K+M)!}} P_K^M(\cos\Theta) . \tag{3.5.13}$$

We note the following relations to the coefficients A_K and B_K of Cambi, Mosconi, and Ricci (CaM 82a)

$$A_K = V_{00}^K \tag{3.5.14}$$

$$B_K = -\sqrt{\frac{(K-2)!}{(K+2)!}} W_{00}^K . \qquad (3.5.15)$$

Corresponding expansions can be obtained for the one- and two-nucleon polarization observables [Are 88a]. They are listed in appendix C.

For the total cross section and its beam and target asymmetries as defined in (3.4.21-23) one finds then

$$\sigma_{tot}^0 = 4\pi V_{00}^0 \qquad (3.5.16)$$

$$\tau_{20}^0 = V_{20}^0 / V_{00}^0 \qquad (3.5.17)$$

$$\tau_{10}^c = V_{10}^0 / V_{00}^0 \qquad (3.5.18)$$

$$\tau_{22}^l = W_{22}^0 / V_{00}^0 . \qquad (3.5.19)$$

4 Methods of calculation for the T-matrix

In the preceding section we have shown how the observables can be expressed in terms of the T-matrix elements or in terms of the multipole amplitudes $\mathcal{O}^{L\lambda}(\mu j l s)$ which contain all the relevant information on the dynamical properties of the system. For the calculation of these amplitudes a variety of different methods have been developed utilizing various kinds of approximations. One may distinguish them according to

(i) relativistic and non-relativistic approaches and intermediate steps by adding relativistic corrections of lowest order to a non-relativistic theory,

(ii) the way how the final-state interaction between the outgoing proton and neutron is treated, whether it is fully taken into account in a coupled-channel approach or only approximately in a diagrammatic expansion,

(iii) whether exchange currents as manifestation of subnuclear degrees of freedom have been included explicitly or not. This distinction, however, is not very clear-cut because many of those approaches, which do not use explicit exchange currents, have partly included them by using the Siegert operators for electric transitions (see section 4.1 below).

Another classification can be made according to the energy region, whether one is considering photon energies below pion production threshold, where only the two-body p-n final channel is possible, or above, where the additional three-body $NN\pi$ channel is present. While below threshold most approaches have used a non-relativistic framework with full consideration of the final state interaction, one encounters above pion threshold mainly the diagrammatic approach. This is not surprising, since there a complete treatment is much more involved compared to the subthreshold region because of the additional open three-body channel. Furthermore, a relativistic treatment is much easier in the diagrammatic method. We will now review these various methods.

4.1 Non-relativistic approach

The starting point is a non-relativistic Hamiltonian of the two-body p-n system

$$\hat{H} = \hat{T} + \hat{V} , \tag{4.1.1}$$

where the interaction is given in the form of a realistic N-N potential. The kinetic energy refers to the intrinsic relative motion, since the center-of-mass motion separates completely from the internal one. As we have discussed already in section 3.2, one obtains in this case for the T-matrix (see (3.2.14) and the additional approximation $E_d = 2M$, which, however, is not necessary)

$$T^{\gamma}_{sm_s\lambda m_d} = -\pi(2\alpha kM/\omega)^{1/2}{}^{(-)}\langle k s m_s|\boldsymbol{\varepsilon}(\lambda)\cdot\boldsymbol{J}(\omega)|1m_d\rangle_{n.r.} \tag{4.1.2}$$

where "$n.r.$" indicates the non-relativistic wave functions.

The initial bound deuteron state has the well known form of S- and D-wave components

$$\langle \boldsymbol{r}|1m_d\rangle = \sum_{l=0,2} \frac{u_l(r)}{r} \langle \Omega|(l1)1m_d\rangle , \tag{4.1.3}$$

where Ω denotes the spherical angles of $\boldsymbol{r}$. The final state is an outgoing scattering solution of the p-n continuum characterized by the relative p-n momentum $\boldsymbol{k}$, spin s and projection m_s onto $\boldsymbol{k}$

$$|\boldsymbol{k}sm_s\rangle^{(-)} = \sum_{\mu j m_j l} \hat{l}(l0sm_s|jm_s)e^{-i\delta^j_\mu} U^j_{ls\mu} D^j_{m_j m_s}(K)|\mu j m_j\rangle , \tag{4.1.4}$$

where K rotates the chosen quantization axis into the direction of the relative momentum $\boldsymbol{k}$. For example, if the incoming-photon momentum $\boldsymbol{\omega}$ is chosen as quantization axis, K would be described by the Euler angles $(0, -\Theta, -\Phi)$ according to fig. 3.2.1.

The partial waves

$$|\mu j m_j\rangle = \sum_{l's'} U^j_{l's'\mu}|\mu(l's')jm_j\rangle \tag{4.1.5}$$

are solutions of a system of coupled equations, where the Blatt-Biedenharn convention (BlB 52) has been used for the definition of the phase shifts δ^j_μ and the mixing parameters ϵ_j, which determine the matrix U in the form

		μ			
l	s	1	2	3	4
$j-1$	1	$\cos\epsilon_j$	0	$-\sin\epsilon_j$	0
j	0	0	1	0	0
$j+1$	1	$\sin\epsilon_j$	0	$\cos\epsilon_j$	0
j	1	0	0	0	1

(4.1.6)

In detail one has

$$\langle \boldsymbol{r}|\mu(ls)jm_j\rangle = \frac{1}{kr}\sqrt{\frac{2}{\pi}} v^j_{ls\mu}(r)\langle\Omega|(ls)jm_j\rangle \tag{4.1.7}$$

with the asymptotic normalization

$$v^j_{ls\mu}(r) \underset{kr\to\infty}{\longrightarrow} \sin(kr - l\pi/2 + \delta^j_\mu) . \tag{4.1.8}$$

The spin-angular functions contain an i^l-factor in order to have the proper transformation under time reversal

$$\langle\Omega|(ls)jm_j\rangle = i^l[Y^{[l]}(\Omega) \times \chi^{[s]}]^{[j]}_{m_j} , \tag{4.1.9}$$

where the two-nucleon spin wave function is denoted by $\chi^{[s]}$. The notation of Fano and Racah (FaR 59) is used for spherical tensors.

For a given N-N potential the radial wave functions $v^j_{ls\mu}(r)$ are then obtained by a numerical solution of the coupled equations and the current operator is evaluated between these non-relativistic wave functions, usually in r-space. In view of the fact that the potentials are fitted to the on-shell phase shifts a large uncertainty remains as to their off-shell behaviour. This has been studied by using phase-equivalent unitary transformations of the Hamiltonian (GrW 83, McT 83). A certain influence on the off-shell behaviour has been found in the total and differential cross section and in particular in the photon asymmetry Σ^l (GrW 83). However, these results should be taken with caution since the electromagnetic current has not been changed even though the non-locality introduced by these transformations is expected to have an influence also on the current in generating an additional two-body contribution.

In this context, we would like to mention a recent approach by von Ferber, Sandhas, and Haberzettl (VFS 89) who have derived a momentum-space formulation in terms of non-singular scattering integral equations. The interesting feature of this approach is the fact that the T-matrix of the final-state interaction enters only on-shell, which can thus be determined completely from experimental phase shifts. Exchange currents are included via the Siegert operators only.

In order to fully exploit the power of angular-momentum techniques one uses the multipole expansion of the current operator

$$\begin{aligned} \boldsymbol{\varepsilon}(\lambda)\cdot\boldsymbol{J}(\boldsymbol{\omega}) &= \int d^3x\,\boldsymbol{\varepsilon}(\lambda)\cdot\hat{\boldsymbol{j}}(\boldsymbol{x})e^{i\boldsymbol{\omega}\cdot\boldsymbol{x}} \\ &= -\sqrt{2\pi}\sum_{LM}\hat{L}(\hat{T}^{[L]}_{e,M}+\lambda\hat{T}^{[L]}_{m,M})\,D^L_{M\lambda}(R)\,, \end{aligned} \tag{4.1.10}$$

where R describes the rotation which carries $\boldsymbol{\omega}$ into the quantization axis usually chosen to be the relative momentum $\boldsymbol{k}$ (see fig. 3.2.1), i.e., $R=(0,-\Theta,-\Phi)$.

The electric and magnetic multipole operators are defined as

$$\hat{T}^{[L]}_{e/m} = \int d^3x\,\hat{\boldsymbol{j}}(\boldsymbol{x})\cdot\boldsymbol{A}^{[L]}(e/m)\,, \tag{4.1.11}$$

where the magnetic (m) and electric (e) multipole fields are given by

$$\begin{aligned} \boldsymbol{A}^{[L]}(m) &= i^L j_L(\omega r)\boldsymbol{Y}^{[L]}_L \\ &= -\frac{i^{L+1}}{\sqrt{L(L+1)}}(\boldsymbol{r}\times\boldsymbol{\nabla})j_L(\omega r)Y^{[L]} \end{aligned} \tag{4.1.12}$$

$$\begin{aligned} \boldsymbol{A}^{[L]}(e) &= \frac{i}{\omega}\boldsymbol{\nabla}\times\boldsymbol{A}^{[L]}(m) \\ &= i^{L+1}\left[\frac{1}{\omega}\sqrt{\frac{L+1}{L}}\boldsymbol{\nabla}(j_L(\omega r)Y^{[L]})-\sqrt{\frac{2L+1}{L}}j_{L+1}(\omega r)\boldsymbol{Y}^{[L]}_{L+1}\right] \\ &= \frac{i^{L+1}}{\sqrt{L(L+1)}}\left[\frac{1}{\omega}\boldsymbol{\nabla}\left(1+r\frac{d}{dr}\right)j_L(\omega r)Y^{[L]}+\omega\boldsymbol{r}j_L(\omega r)Y^{[L]}\right]\,. \end{aligned} \tag{4.1.13}$$

The spherical harmonics are denoted by $Y^{[L]}$. For the electric multipole fields two of the many possible forms are given explicitly which allow to separate the low-energy limit of the Siegert theorem. Recently, Friar and Fallieros have proposed another form in order to optimize the Siegert operators (FrF 84).

We will now shortly review the Siegert theorem (Sie 37). The starting point is the general separation

$$\boldsymbol{A}^{[L]}(e) = \frac{i^{L+1}}{\omega}\boldsymbol{\nabla}(\Phi_L(\omega r)Y^{[L]})+\boldsymbol{A}'^{[L]}(e)\,, \tag{4.1.14}$$

where Φ_L may be chosen arbitrarily except for the low-energy limit

$$\Phi_L(\omega r)/\omega^L \underset{\omega\to 0}{\longrightarrow} \sqrt{\frac{L+1}{L}} \frac{r^L}{(2L+1)!!}, \tag{4.1.15}$$

$$A'^{[L]}(e)/\omega^L \underset{\omega\to 0}{\longrightarrow} 0 . \tag{4.1.16}$$

Then with the help of current conservation

$$\nabla \cdot \hat{\boldsymbol{j}}(\boldsymbol{x}) + i[\hat{H}, \hat{\rho}(\boldsymbol{x})] = 0 \tag{4.1.17}$$

one obtains for the electric multipole moments

$$\hat{T}_e^{[L]} = -\frac{i^L}{\omega}[H, \int d^3x \hat{\rho}(x) \Phi_L(\omega x) Y^{[L]}] + \int d^3x \hat{\boldsymbol{j}}(\boldsymbol{x}) \cdot \boldsymbol{A}'^{[L]}(e) . \tag{4.1.18}$$

The first part on the rhs, i.e., the commutator with the Hamiltonian is the so-called Siegert operator, which gives the dominant contribution at low energies.

As remarked above this operator refers to intrinsic variables only, i.e., $\hat{H}$ is the intrinsic Hamiltonian and $\hat{\boldsymbol{j}}(\boldsymbol{x})$ the intrinsic current. Thus one obtains for the transition matrix element

$$\begin{aligned} \langle f|\hat{T}_e^{[L]}|d\rangle &= -\frac{i^L}{\omega}(E_f+\epsilon)\int d^3x \langle f|\hat{\rho}(x)|d\rangle \Phi_L Y^{[L]} \\ &\quad + \int d^3x \langle f|\hat{\boldsymbol{j}}(\boldsymbol{x})|d\rangle \cdot \boldsymbol{A}'^{[L]}(e) , \end{aligned} \tag{4.1.19}$$

where E_f is the final state energy, i.e., $\hat{H}|f\rangle = E_f|f\rangle$. Taking $E_f = E_{np}$ from (3.1.8) one finds

$$\begin{aligned} \frac{1}{\omega}(E_f+\epsilon) &= \frac{1}{\omega}(\omega + E_d - 2M + \epsilon) \\ &= \frac{1}{\omega}\left(\omega + M_d + \frac{1}{2}\frac{\omega^2}{M_d} - M_d\right) \\ &\approx 1 + \frac{\omega}{2M_d} \end{aligned} \tag{4.1.20}$$

as has been used by Partovi (Par 64). The additional term $\omega/2M_d$ reflects the fact that the internal excitation energy is not only supplied by the photon but also by the kinetic energy of the incoming deuteron in the overall cm system.

However, we would like to remark that the use of (4.1.20) is legitimate only if E_{np} is indeed the energy of the final n-p state, i.e., eigenvalue of the intrinsic Hamiltonian. But that is not the case because of the inconsistency of using relativistic kinematics and non-relativistic dynamics. In fact, the final-state energy is determined by the asymptotic relative momentum $\boldsymbol{k}$ as given in (3.1.12). From there one finds

$$\begin{aligned} E_f &= \frac{\boldsymbol{k}^2}{M} \\ &= -\epsilon + \frac{\epsilon^2}{4M} + \frac{\omega}{2M}(\omega + E_d) \\ &\approx -\epsilon + \omega\left(1 + \frac{\omega}{2M}\right) \end{aligned} \tag{4.1.21}$$

and thus

$$\frac{1}{\omega}(E_f+\epsilon)\approx\left(1+\frac{\omega}{2M}\right) . \tag{4.1.22}$$

The difference to (4.1.20) seems to be small but leads already at 100 MeV to a 50 percent difference in the total cross section. Furthermore, (4.1.20) introduces a slight gauge dependence even for a gauge-invariant current in contrast to (4.1.21).

A second remark is concerned with the inclusion of the cm current by Jaus and Woolcock (JaW 84a, JaW 87a) who have replaced (4.1.20) by one. But this is not correct in the non-relativistic limit, since the effective current which includes the cm current does not obey the current-conservation relation (4.1.17) with the total Hamiltonian including the cm motion (MoR 87, Are 88b, MoR 88).

The important feature of the relation (4.1.19) is that it allows to evaluate the dominant part of the electric transitions without explicit knowledge of the nuclear current density. Only the charge density is needed. However, for the remaining part of (4.1.19) and for the magnetic transitions knowledge of the explicit form of the current is necessary.

From the multipole expansion and the wave functions as described above one obtains the reduced multipole amplitudes which have been introduced in (3.5.1) in the form

$$\mathcal{O}^{L\lambda}(\mu jls) = (4\pi)^{1/2}e^{i\delta^j_\mu}U^j_{ls\mu}(E^L_{\mu j}+\lambda M^L_{\mu j}) \tag{4.1.23}$$

introducing

$$E^L_{\mu j} = \langle \mu j \parallel \hat{T}^{[L]}_e \parallel 1\rangle \tag{4.1.24}$$

for the electric transitions and

$$M^L_{\mu j} = \langle \mu j \parallel \hat{T}^{[L]}_m \parallel 1\rangle \tag{4.1.25}$$

for the magnetic transitions, which correspond to $\mathcal{J}^{(L)}(\mu j)$ and $\mathcal{S}^{(L)}(\mu j)$, respectively, of Partovi (Par 64). The occurrence of the strong interaction phase shifts in (4.1.23) is a manifestation of Watson's final-state theorem (Wat 54).

For the calculation of the T-matrix one now has to specify the charge and current density operators. Within the non-relativistic theory one separates them into one- and many-body operators

$$\hat{\rho}(\boldsymbol{x}) = \hat{\rho}_{(1)}(\boldsymbol{x})+\hat{\rho}_{(2)}(\boldsymbol{x})+\cdots \tag{4.1.26}$$

$$\hat{\boldsymbol{\jmath}}(\boldsymbol{x}) = \hat{\boldsymbol{\jmath}}_{(1)}(\boldsymbol{x})+\hat{\boldsymbol{\jmath}}_{(2)}(\boldsymbol{x})+\cdots , \tag{4.1.27}$$

where the one-body parts are given by the conventional non-interacting point-particle expressions

$$\hat{\rho}_{(1)}(\boldsymbol{x}) = \sum_l e_l\delta(\boldsymbol{x}-\boldsymbol{r}_l) \tag{4.1.28}$$

$$\hat{\boldsymbol{\jmath}}_{(1)}(\boldsymbol{x}) = \frac{1}{2M}\sum_l\left(e_l\{\boldsymbol{p}_l,\delta(\boldsymbol{x}-\boldsymbol{r}_l)\}+i\mu_l\boldsymbol{\sigma}(l)\times[\boldsymbol{p}_l,\delta(\boldsymbol{x}-\boldsymbol{r}_l)]\right) . \tag{4.1.29}$$

Here e_l and μ_l denote charge and magnetic moment of the l-th particle having coordinate $\boldsymbol{r}_l$, momentum $\boldsymbol{p}_l$ and spin operator $\boldsymbol{\sigma}(l)$. The many-body operators, the so-called exchange operators, arise in principle from the interaction between the particles. They are, however, not completely independent. Because if one separates the continuity equation (4.1.17) into one- and many-body operators, one finds up to two-body contributions

$$\nabla \cdot \hat{\boldsymbol{\jmath}}_{(1)}(\boldsymbol{x}) + i[\hat{T}, \hat{\rho}_{(1)}(\boldsymbol{x})] = 0 \tag{4.1.30}$$

$$\nabla \cdot \hat{\boldsymbol{\jmath}}_{(2)}(\boldsymbol{x}) + i[\hat{T}, \hat{\rho}_{(2)}(\boldsymbol{x})] + i[\hat{V}, \hat{\rho}_{(1)}(\boldsymbol{x})] = 0\,. \tag{4.1.31}$$

Assuming that the two-body exchange charge density $\hat{\rho}_{(2)}$ vanishes in the non-relativistic limit (Siegert hypothesis) which in fact is valid for explicit meson-exchange models in the static limit, then, according to (4.1.31), one finds the necessity to consider two-body exchange currents if the potential does not commute with the one-body charge density. Since (4.1.31) is not sufficient to determine $\hat{\boldsymbol{\jmath}}(2)$ uniquely, one needs a specific dynamic model consistent with the interaction potential.

It is now evident from (4.1.30) and (4.1.31) that the Siegert operator of (4.1.18) includes the exchange current contribution via the potential part of the commutator (FaA 76, Lag 78, HwM 80, Are 81c, GaH 81). Therefore, one has to keep in mind this fact, when one considers calculations where no explicit exchange currents have been included but use of the Siegert operators has been made. In fact, what is called the classical theory takes as explicit current the one-body current (4.1.29) only but uses for the electric transitions the Siegert operators of (4.1.18) and thus includes part of the exchange currents, in fact the dominant part. A step beyond this classical theory is taken by considering explicitly meson-exchange currents or subnuclear degrees of freedom.

4.2 Meson-exchange currents and isobar configurations

As mentioned in the previous section all realistic N-N interaction models are isospin- and/or momentum-dependent and, therefore, additional interaction currents are required in order to satisfy current conservation (4.1.31). It is, however, a well-known fact that the gauge condition for the current is not sufficient to determine uniquely the exchange current because it constrains the longitudinal current only. Thus an arbitrary transverse current can be added without destroying the gauge condition. For this reason one needs an explicit dynamical model from which not only the N-N interaction but also the corresponding exchange currents are uniquely obtained as effective operators (ChR 71, GaH 76, Che 79, Mat 89, Ris 89).

Within the conventional framework of nuclear physics the coupling of nucleon to meson and isobar degrees of freedom provides such a dynamical model which has been proven to be quite successful. The meson degrees of freedom are usually eliminated by introducing suitable effective operators which take their presence into account implicitly.

By standard methods one finds, for example, for the one-pion-exchange interaction (pseudoscalar-isovector exchange)

$$\begin{aligned}\hat{V}^{OPE} &= \frac{f^2}{m^2}\boldsymbol{\tau}_1\cdot\boldsymbol{\tau}_2[\boldsymbol{\sigma}_1\cdot\boldsymbol{p}_1,[\boldsymbol{\sigma}_2\cdot\boldsymbol{p}_2, J_m(\boldsymbol{r}_{12}) - J_\Lambda(\boldsymbol{r}_{12})]]\\ &= -\frac{f^2}{m^2}\boldsymbol{\tau}_1\cdot\boldsymbol{\tau}_2[\boldsymbol{\sigma}_1\cdot\nabla_1)(\boldsymbol{\sigma}_2\cdot\nabla_2)(J_m(\boldsymbol{r}_{12}) - J_\Lambda(\boldsymbol{r}_{12}))\end{aligned} \tag{4.2.1}$$

with

$$J_\alpha(\boldsymbol{r}) = \frac{1}{(2\pi)^3}\int d^3q \frac{e^{i\boldsymbol{q}\cdot\boldsymbol{r}}}{\boldsymbol{q}^2+\alpha^2} = \frac{e^{-\alpha r}}{4\pi r}\,, \tag{4.2.2}$$

where $\boldsymbol{\sigma}$ and $\boldsymbol{\tau}$ denote the nucleon spin and isospin operators, m the pion mass and f the πNN coupling constant. Here we have assumed a monopole regularization, i.e., a πNN vertex form factor of the form

$$g(\boldsymbol{q}) = \sqrt{\frac{\Lambda^2 - m^2}{\Lambda^2 + \boldsymbol{q}^2}}\,, \tag{4.2.3}$$

where its range is characterized by the regularization parameter Λ.

Without hadronic form factors the two-body meson-exchange current can be obtained from the non-relativistic reduction of the corresponding Feynman diagram (ChR 71). But it is well known that in the presence of hadronic π-N form factors this method does not yield a gauge-invariant current. Because the existence of an extended or dressed vertex is related to the internal structure of the participating particles which in turn manifests itself in self-energy contributions. In this case additional terms arise from the self-energy contributions in order to fulfil the general Ward-Takahashi identity. A further complication arises for off-shell particles, because additional off-shell form factors will appear in principle, which are, however, not known. But in a phenomenological manner a gauge-invariant two-body exchange current can still be obtained for a regularized boson-exchange potential by minimal coupling (Are 79, Sch 86, ScA 89a, Oht 89).

For example, for the monopole regularized V^{OPE} of (4.2.1) it has the form

$$\begin{aligned}\hat{\boldsymbol{\jmath}}^{OPE}_{(2)}(\boldsymbol{x},\boldsymbol{r}_1,\boldsymbol{r}_2) = {} & \frac{f^2}{m^2}(\boldsymbol{\tau}_1 \times \boldsymbol{\tau}_2)_3 \\ & \cdot \big(\delta(\boldsymbol{x}-\boldsymbol{r}_2)\boldsymbol{\sigma}_2(\boldsymbol{\sigma}_1 \cdot \boldsymbol{\nabla}_1)(J_m(\boldsymbol{r}_{12}) - J_\Lambda(\boldsymbol{r}_{12})) - (1 \leftrightarrow 2) \\ & - \boldsymbol{\sigma}_1 \cdot \boldsymbol{\nabla}_1 \boldsymbol{\sigma}_2 \cdot \boldsymbol{\nabla}_2 (J_m(\boldsymbol{r}_1 - \boldsymbol{x}) \overset{\leftrightarrow}{\underset{x}{\boldsymbol{\nabla}}} J_m(\boldsymbol{x}-\boldsymbol{r}_2) \\ & - J_\Lambda(\boldsymbol{r}_1 - \boldsymbol{x}) \overset{\leftrightarrow}{\underset{x}{\boldsymbol{\nabla}}} J_\Lambda(\boldsymbol{x}-\boldsymbol{r}_2))\big)\,,\end{aligned} \tag{4.2.4}$$

where the first part describes the so-called pair current related to the momentum dependence of the πNN vertex at low momentum transfer. The second part describes the interaction of a photon with the exchanged pion, the pionic current (see fig. 4.2.1). It is a simple exercise to show that this current indeed fulfills the gauge condition

$$\boldsymbol{\nabla} \cdot \hat{\boldsymbol{\jmath}}^{OPE}_{(2)} + i[\hat{V}^{OPE}, \hat{\rho}_{(1)}] = 0 \tag{4.2.5}$$

with the one-body charge density (4.1.28). The expressions for higher-order regularization can be obtained by differentiation with respect to Λ^2. Analogously, the ρ-exchange current can be constructed for the corresponding diagrams in fig. 4.2.1 where the pion is replaced by the ρ-meson with a corresponding change of the vertex structure.

In this way a completely consistent exchange current can be constructed from an underlying meson-nucleon field-theoretical model. If, however, the N-N interaction model is to a large extent of phenomenological nature the situation is less clear-cut but still not hopeless concerning the construction of a decent exchange current. The reason for this is that most of the phenomenological potential models use some ingredients of an underlying meson-nucleon picture. Thus all have in the long-range part the one-pion-exchange contribution, for which the corresponding exchange current is well known. Uncertainties arise then in the intermediate- and short-range part. But even then it is sometimes possible to construct a consistent even though not unique exchange current,

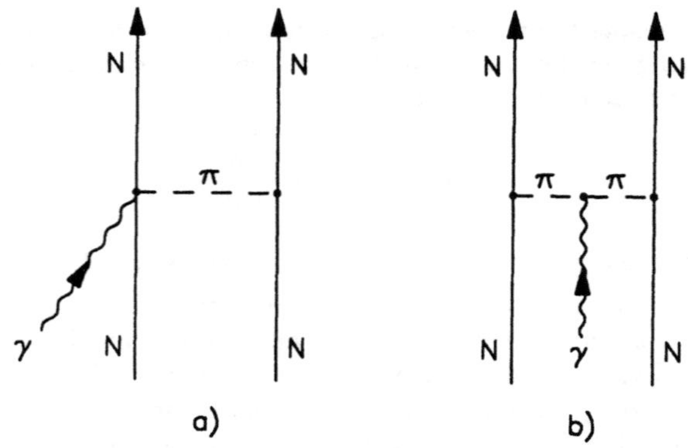

Fig. 4.2.1: Pion-exchange-current diagrams: (a) contact or pair current, (b) pion current.

which relies on meson-theoretical background as has been demonstrated for the Paris potential (BuL 85, Ris 85). Similar methods for constructing gauge-invariant two-body currents for a given phenomenological potential are proposed in LvL 87 and Par 87 which also rely on the minimal subsitution.

Strong criticism has been raised recently in NaK 86 as to the consistency of exchange currents in non-relativistic theories claiming that uncontrollable uncertainties are introduced by the violation of gauge invariance. However, it has been shown in ScW 90 that this criticism is unfounded and that no severe violation of gauge invariance occurs.

In the evaluation of the exchange-current contribution to the electric matrix elements one has to be careful if the Siegert operators (see (4.1.18)) have been used. Because in this case a large part of the MEC contributions is already included as explained above and thus one has to subtract this contribution, the so-called Siegert correction (FaA 76). That seems to have been overlooked in ReM 67, where MEC contributions to electric transitions have been considered for the first time and found to lead to a strong reduction of the total cross section below $E_\gamma = 60$ MeV in contrast to later findings of other authors.

Another important contribution arises from the internal nucleon degrees of freedom as manifest in virtual excitations of isobars in the nuclear medium. Examples of their contributions to the N-N interaction and the electromagnetic current are shown in fig. 4.2.2. Again one can eliminate these internal nucleon degrees of freedom in the standard way by introducing appropriate effective operators. For example, one obtains for the N-N interaction a dispersive contribution from the diagram of fig. 4.2.3.

$$V_{NN}^{disp} = -\sum_{N_1,N_2} V_{NN,N_1N_2}(H_{N_1N_2} - E)^{-1} V_{N_1N_2,NN} \tag{4.2.6}$$

which leads to intermediate-range attraction.

Correspondingly one finds a current contribution from the diagrams of fig. 4.2.4

$$\boldsymbol{j}^{(\Delta)} = -V_{NN,N\Delta}(H_{\Delta N} - E)^{-1} \boldsymbol{j}_{\Delta N} + h.c. \,, \tag{4.2.7}$$

where $\boldsymbol{j}_{\Delta N}$ is the ΔN transition current describing the electromagnetic Δ excitation, which has the form (WeA 78)

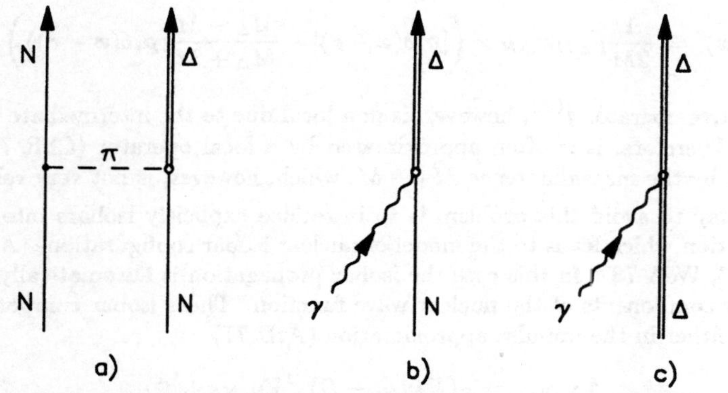

Fig. 4.2.2: Diagrams of Δ-contribution to two-body interaction (a) and electromagnetic current (b, c).

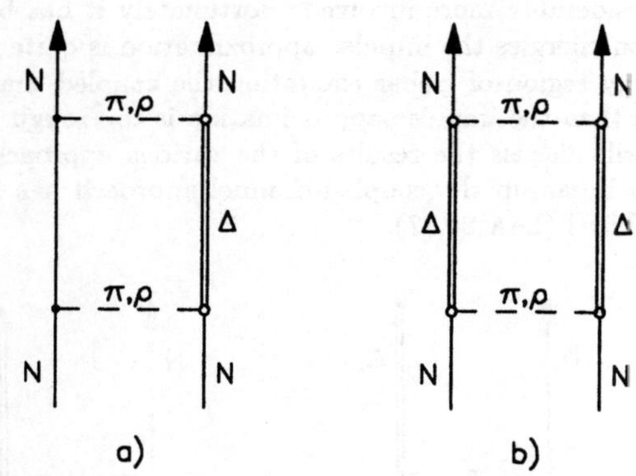

Fig. 4.2.3: Diagrams of dispersive contribution from intermediate Δ (a) and $\Delta\Delta$ (b) excitation to effective N-N potential.

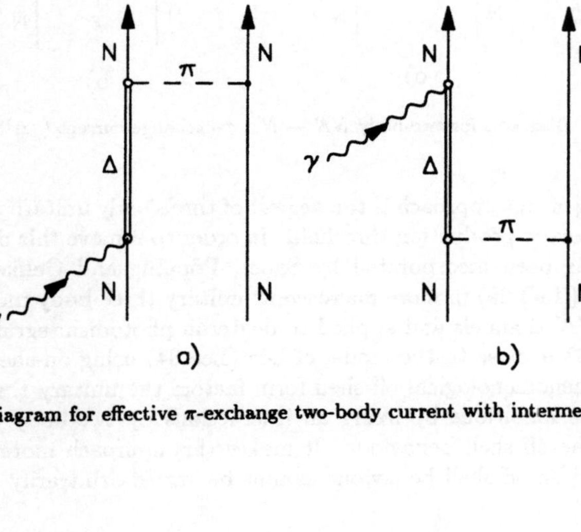

Fig. 4.2.4: Diagram for effective π-exchange two-body current with intermediate Δ excitation.

$$\boldsymbol{j}_{\Delta N}(\boldsymbol{x}) = \frac{1}{2M}\mu_{\Delta N}\boldsymbol{\sigma}_{\Delta N} \times \left([\boldsymbol{p}, \delta(\boldsymbol{x}-\boldsymbol{r})] - \frac{M_\Delta - M}{M_\Delta + M}\{\boldsymbol{p}, \delta(\boldsymbol{x}-\boldsymbol{r})\}\right) . \quad (4.2.8)$$

This effective operator $\boldsymbol{j}^{(\Delta)}$, however, is non-local due to the intermediate isobar propagation. Therefore, it is often approximated by a local operator (ChR 71) replacing $H_{\Delta N} - E$ by the mass difference $M_\Delta - M$, which, however, is not very reliable.

One way to avoid this problem is to introduce explicitly isobars into the nuclear wave function which leads to the model of nuclear isobar configurations (ArD 71, ArW 72, Gre 76, WeA 78). In this case the isobar propagation is automatically included in the isobar components of the nuclear wave function. These isobar components can be obtained either in the impulse approximation (ArD 71)

$$\Phi_{N_1N_2} = -(H_{N_1N_2} - E)^{-1}V_{N_1N_2,NN}\Phi_{NN} , \quad (4.2.9)$$

if the normal nucleonic wave function is known, or by solving directly the resulting coupled equations (GrS 79, LeA 84,87, VFT 84). The latter method is certainly more accurate but also considerably more involved. Fortunately it has been found that at least for low excitation energies the impulse approximation is quite reliable. However, approaching the energy region of isobar excitation the coupled-channel method is to be preferred, because then the impulse approximation is too rough. This will become apparent, when we will discuss the results of the various approaches. For the electromagnetic deuteron break-up the coupled-channel approach has first been used by Leidemann and Arenhövel (LeA 84,87).

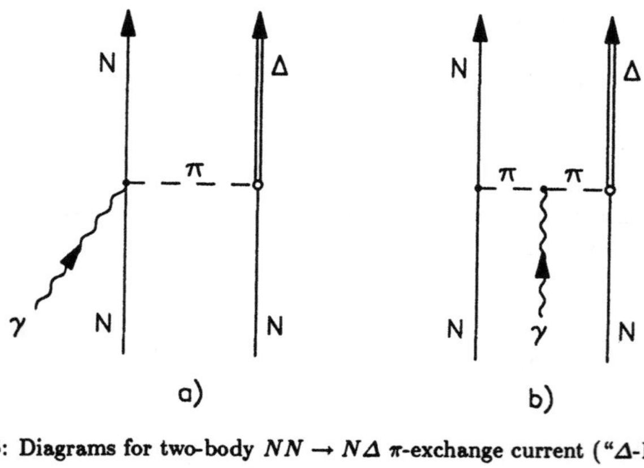

Fig. 4.2.5: Diagrams for two-body $NN \to N\Delta$ π-exchange current ("Δ-MEC").

A shortcoming of this approach is the neglect of three-body unitarity due to explicit πNN channels above π-production threshold. In order to remove this deficiency three-body unitarity has been incorporated by Sauer, Pöpping and Oelfke (Sau 87) and Tanabe and Ohta (TaO 89) in more microscopic unitary three-body models of coupled NN, $N\Delta$ and πNN channels and applied to deuteron photodisintegration. While the approach of Sau 87 is close to the model of Lee (Lee 84) using on-shell πN and NN t-matrices with phenomenological off-shell form factors the unitary three-body model of TaO 89 is more ambitious by fixing all interactions by two-body scattering data which fixes also the off-shell behaviour. It makes this approach more consistent but less flexible, since the off-shell behaviour cannot be varied arbitrarily as in the other approaches. In fact the phase shifts of the low partial waves are poorly described, in particular for the ^3P_j-waves. In order to improve the results the Paris potential is also used in TaO 89. Furthermore, the deuteron wave function is taken from the Paris model too, thus loosing consistency.

Finally, we would like to mention MEC contributions which involve isobar configurations as illustrated by the diagrams in fig. 4.2.5. They describe the π- exchange $NN \rightarrow N\Delta$ transition current. The corresponding operators can be derived in a completely analogous manner like the π-MEC operators. Their importance in deuteron photo- and electrodisintegration near threshold has first been demonstrated by Leidemann and Arenhövel (Lei 80, Are 82, LeA 83). They have also been considered in GrS 82b for thermal n-p capture. In the later discussion we will use "Δ-MEC" as a shorthand for these contributions.

4.3 Relativistic effects

For a long time relativistic effects have been considered unimportant for deuteron photodisintegration below pion threshold. Since the deuteron is a loosely bound object of large spatial extension, the wave function is dominated by low momentum components. Only at higher energies, in the region of the Δ isobar and above, covariant approaches or Feynman diagram techniques have been used in order to include expected relativistic effects. Therefore it came quite as a surprise when Cambi, Mosconi and Ricci (CaM 82b) demonstrated the importance of the relativistic spin-orbit contribution to the charge density for the 0°-cross section even at energies as low as 20 MeV. The reason for this is a specific kinematic situation because at these low energies the 0°-cross section is about an order of magnitude smaller than at 90° due to a destructive interference of the dominant $E1$ transitions to the ^3P_j final states. Therefore, any relatively small correction compared to the size of the individual matrix elements can have a relatively large influence on the 0°-cross section as in fact is the case for the relativistic correction given by the spin-orbit current. A more detailed account can be found in section 7.3.

Since then more systematic studies of relativistic effects in the photodisintegration of the deuteron have been done. One may distinguish essentially three classes of relativistic effects which, however, are interrelated and thus require a consistent treatment (Fri 79). First, one has to deal with relativistic effects in the wave function of an interacting system in its rest frame, that means relativistic effects in the internal dynamics. Second, one has to boost the wave function to a moving frame which not only affects the spin degrees of freedom but the internal structure as well (Lorentz contraction). Finally, also the current operator has to be supplemented by terms of relativistic order. One important consistency check will be the gauge condition for the current which has to be fulfilled up to the order considered. The other will be the covariance of the current and charge matrix elements or at least approximate covariance up to the corresponding order.

The usual way to incorporate relativistic effects is a p/M-expansion of a relativistic model, where for the nuclear potential V/M is considered as order $(p/M)^2$. Normally one contents oneself with the lowest-order correction of $(p/M)^2$ beyond the nonrelativistic operators. Two approaches have been used. One is a Foldy-Wouthuysen reduction of a Dirac nucleon coupled to external potentials or fields, while the other starts from a covariant Blankenbecler-Sugar reduction of the Bethe-Salpeter equation for two interacting nucleons, also called quasi-potential approach (JaW 84b).

The first method has been used by Cambi, Mosconi and Ricci (CaM 82b, 84) and by Friar, Gibson and Payne (FrG 84). The point of departure is the Dirac equation for a nucleon

$$((\not p - e\not A - \not V_v) - (m + V_s) - \left[\left((1-\mu)ig_\pi - \frac{\mu f_\pi}{m_\pi}\not\partial\right)\boldsymbol{\tau}\cdot\boldsymbol{\Phi}_\pi\right]\Psi = 0 \qquad (4.3.1)$$

which is coupled to electromagnetic and phenomenological vector and scalar and pion fields, here in a mixture of *ps* and *pv* coupling ($0 < \mu < 1$). In the following we will set $f_\pi/m_\pi = g_\pi/2M$. A Foldy-Wouthuysen reduction and elimination of the other fields by introducing effective operators gives as Hamiltonian for two interacting nucleons including relativistic corrections

$$H(1,2) = H_0(1,2) + H_{RC}(1,2) \qquad (4.3.2)$$

$$H_0(1,2) = \frac{1}{2M}(p_1^2 + p_2^2) + V_s + V_v - \left(\frac{g_\pi}{2M}\right)^2 \boldsymbol{\tau}_1\cdot\boldsymbol{\tau}_2\boldsymbol{\sigma}_1\cdot\boldsymbol{\nabla}_1\boldsymbol{\sigma}_2\cdot\boldsymbol{\nabla}_2 Y_{(2)} \qquad (4.3.3)$$

$$\begin{aligned} H_{RC}(1,2) = & -\frac{p_1^4}{8M^3} + \frac{1}{8M^2}\nabla_1^2(V_v - V_s) - \frac{1}{4M^2}\{p_1^2, V_s\} \\ & + \frac{1}{8M^3}\{\boldsymbol{p}_1; \boldsymbol{\sigma}_1 \times \boldsymbol{\nabla}_1(V_v - V_s)\} \\ & + \frac{1}{8M^2}\left(\frac{g_\pi}{2M}\right)^2 \boldsymbol{\tau}_1\cdot\boldsymbol{\tau}_2\{\{p_1^2 + p_2^2, \boldsymbol{\sigma}_1\cdot\boldsymbol{\nabla}_1\boldsymbol{\sigma}_2\cdot\boldsymbol{\nabla}_2 Y_{(2)}\} \qquad (4.3.4) \\ & - i\mu[p_1^2, \{\boldsymbol{\sigma}_1\cdot\boldsymbol{p}_1, \boldsymbol{\sigma}_2\cdot\boldsymbol{\nabla}_2 Y_{(2)}\}] \\ & - \frac{i}{2}(1+\mu)[p_2^2 - p_1^2, \{\boldsymbol{\sigma}_1\cdot\boldsymbol{p}_1, \boldsymbol{\sigma}_2\cdot\boldsymbol{\nabla}_2 Y_{(2)}\}] \\ & - [p_1^2, [p_1^2, \boldsymbol{\sigma}_1\cdot\boldsymbol{\nabla}_1\boldsymbol{\sigma}_2\cdot\boldsymbol{\nabla}_2 Y_{(4)}]]\} + (1 \leftrightarrow 2) \end{aligned}$$

and corresponding charge and current operators of which we list here explicitly only the relativistic corrections of Darwin-Foldy and spin-orbit type

$$\begin{aligned} \rho_{RC}(\boldsymbol{x}) = & \frac{1}{8M^2}(2\mu_1 - e_1)(\nabla_1^2\delta(\boldsymbol{x} - \boldsymbol{r}_1) - \{\boldsymbol{\sigma}_1 \times \boldsymbol{p}_1, \boldsymbol{\nabla}_1\delta(\boldsymbol{x} - \boldsymbol{r}_1)\}) + (1 \leftrightarrow 2) \\ & + \cdots, \end{aligned} \qquad (4.3.5)$$

$$\begin{aligned} \boldsymbol{j}_{RC}(\boldsymbol{x}) = & \frac{i}{8M^2}(2\mu_1 - e_1)[H_0, \boldsymbol{\nabla}_1\delta(\boldsymbol{x} - \boldsymbol{r}_1) - \{\boldsymbol{\sigma}_1 \times \boldsymbol{p}_1, \delta(\boldsymbol{x} - \boldsymbol{r}_1)\}] + (1 \leftrightarrow 2) \\ & + \cdots . \end{aligned} \qquad (4.3.6)$$

We have introduced in (4.3.3-4)

$$Y_{(n)}(\boldsymbol{r}) = \frac{1}{(2\pi)^3}\int d^3q\, g^2(q^2)\frac{e^{i\boldsymbol{q}\cdot\boldsymbol{r}}}{(q^2 + m_\pi^2)^{n/2}} \qquad (4.3.7)$$

where $g(q^2)$ denotes a hadronic πNN vertex form factor.

The first part in (4.3.5-6) is the well-known Darwin-Foldy correction to charge and current density. Since its current is purely longitudinal it will not contribute to photodisintegration. The second part is called the spin-orbit charge and current density. Its current contains a transverse piece which as it turns out gives the most important relativistic effect (CaM 82b) and which arises essentially from its two-body part as given by the potential contribution to the commutator in (4.3.6) (WiL 88) (see section 7.3).

The relativistic two-body corrections contain pieces which arise from retardation or non-static effects in the N-N interaction and associated currents as mediated by meson exchange. These retardation corrections are required by the principles of relativity which forbids instantaneous interactions even though many realistic potentials are based on static, i.e., instantaneous meson exchange. They are easily incorporated in the diagrammatic approach (see section 4.4). Within the non-relativistic approach retardation in the meson-exchange operators has been studied by Hwang and Walker (HwW 85) above π-threshold up to $E_\gamma = 300$ MeV, but retardation in the interaction has been neglected. Recently, retardation effects have been studied consistently by Schmitt and Arenhövel (Sch 86, ScA 89a) applying non-covariant perturbation methods to a simple regularized one-pion-exchange model and to a modified Bryan-Scott potential (BrS 69), where retarded propagators for the exchanged mesons appear explicitly. Subsequently, these investigations have been extended in ScA 89b to the OBEPT approximation of the recent full model of the Bonn potential (MaH 88).

Now the question remains of how to boost the wave function. In the non- relativistic case the intrinsic wave function $\Psi_{d,\boldsymbol{P}}$ of a deuteron moving with momentum $\boldsymbol{P}$ is simply given by the intrinsic rest-frame wave function Ψ_d, i.e.,

$$\Psi_{d,\boldsymbol{P}} = \Psi_d \,. \tag{4.3.8}$$

This is not any more correct if relativistic effects are considered because then one has to include the effects of Lorentz contraction and spin precession. In fact, following the work of Foldy and others, as has been reviewed by Friar (Fri 79), one can obtain the boosted intrinsic wave function including $(v/c)^2$ corrections by

$$\Psi_{d,\boldsymbol{P}} = (1 - i\chi(\boldsymbol{P}))\Psi_d \,, \tag{4.3.9}$$

where

$$\begin{aligned}\chi(\boldsymbol{P}) = & -\frac{1}{4M_d^2}\boldsymbol{r}\cdot\boldsymbol{P}\boldsymbol{p}\cdot\boldsymbol{P} + h.c. \\ & + \frac{1}{4MM_d}((\boldsymbol{\sigma}_1 - \boldsymbol{\sigma}_2)\times\boldsymbol{p})\cdot\boldsymbol{P} + \chi_v(\boldsymbol{P}) \,.\end{aligned} \tag{4.3.10}$$

Here, χ_v describes the potential-dependent part of the boost operator (Fri 77, 79). Since χ is hermitian, one can write this also in the form of the unitary transformation or boost operator as defined in (3.2.9) with

$$U(\boldsymbol{P}) = e^{-i\chi(\boldsymbol{P})} \,. \tag{4.3.11}$$

This allows to retain the cm-frame wave functions by incorporating the boosting effects in the operators

$$\hat{\Omega} \rightarrow \hat{\Omega} + i[\chi, \hat{\Omega}] \,. \tag{4.3.12}$$

Jaus and Woolcock (JaW 81, 84a, 87a,b) have calculated relativistic corrections in a relativistic quasipotential formalism which essentially consists of a Blankenbecler-Sugar (BbS) reduction of the Bethe-Salpeter equation to an equal-time equation (BlS 66). The covariant wave function $\Phi(\boldsymbol{P};\boldsymbol{p})$ for the bound state moving with an arbitrary total momentum $\boldsymbol{P}$ obeys the equation

$$\frac{1}{4M}(M_d^2 - 4M^2 + 4\hat{p}^2)\Phi(\boldsymbol{P};\boldsymbol{p}) = \frac{1}{(2\pi)^3}\int d^3k\tilde{V}(P;\boldsymbol{p},\boldsymbol{k})\Phi(\boldsymbol{P};\boldsymbol{k}) \,, \tag{4.3.13}$$

where

$$\hat{p} = \left(\frac{\boldsymbol{P}\cdot\boldsymbol{p}}{P_0}, \boldsymbol{p}\right) . \tag{4.3.14}$$

For the definition of the quasipotential $\tilde{V}$ we refer to JaW 84b. Due to the energy dependence of $\tilde{V}$ one has a different normalization condition compared to the wave function for an energy-independent potential. Relativistic corrections to the moving bound state with respect to the rest-frame wave function $\Phi_0(\boldsymbol{p}) = \Phi(\boldsymbol{O},\boldsymbol{p})$ are obtained by a Taylor expansion

$$\sqrt{\frac{P_0}{M_d}}\Phi(\boldsymbol{P};\boldsymbol{p}) = \int d^3x e^{-i\boldsymbol{p}\cdot\boldsymbol{x}} \left(1 + \frac{1}{8M^2}(\boldsymbol{P}^2 + \boldsymbol{P}\cdot\boldsymbol{x}\boldsymbol{P}\cdot\nabla - \boldsymbol{P}\cdot(\boldsymbol{\sigma}_1 - \boldsymbol{\sigma}_2)\times\nabla) + O\left(\frac{1}{M^4}\right)\right)\Phi_0(\boldsymbol{x}) . \tag{4.3.15}$$

One should note that the corrections of order M^{-2} are independent of the potential and correspond to the one in (4.3.10). The spin-dependent part describes the Wigner rotation. Relativistic corrections to the charge and current densities are then calculated in an analogous way by applying the BbS reduction to the general photon vertex function of the Bethe-Salpeter formalism and making a power-series expansion in M^{-2} (JaW 84a, 87a).

Finally, we would like to mention the work of Hwang and collaborators (HwM 80, HwL 83, Hwa 84, HwW 85) where also relativistic effects are incorporated in a so-called "elementary-particle treatment". Their results, however, differ substantially from the standard results as obtained by the methods described before. Since the relation of this approach to the standard one has not been discussed by the authors the origin of this discrepancy is unknown, whether it comes from a different dynamical treatment or simply from some numerical deficiency.

4.4 The diagrammatic approach

This approach uses techniques developed in high-energy physics where one contents oneself in considering a limited set of diagrams for a given process starting from lowest order in the hope that these are the leading ones. In fact this method has originally been applied to photodisintegration at high energies above pion production threshold and is also used in the covariant approach as will be discussed in the next section. In the early stages (Aus 55, Wil 56, Zac 56) the calculations were rather crude, allowing only a qualitative description of the gross features. But in recent years it has reached a remarkably high degree of sophistication by including more and more complicated diagrams in the work of Laget (Lag 78), Ogawa et al. (OgK 80), Anastasio and Chemtob (AnC 81) and Levchook (Lev 88) and it gives even at low energies quite satisfactory results (Lag 84).

As an example of such a set of diagrams we show in fig. 4.4.1 the one which has been used in Laget's work (Lag 78, 84). It contains deuteron and nucleon pole diagrams, N-N rescattering terms, pion-exchange-current contribution and isobar-excitation diagrams. The contributions of the various diagrams are either evaluated non-relativistically from the beginning or by reducing the corresponding Feynman diagrams to the non-relativistic order. The latter method allows also to include relativistic corrections. Since the individual diagrams are in general not gauge invariant, care has to be taken to include all those diagrams necessary for gauge invariance, as has been discussed in detail in Lag 78. It is also worth noting that no Siegert operators appear since no multipole decomposition is done and thus the full MEC contribution is exhibited explicitly.

An advantage of this approach and also of the covariant analysis discussed below is that no truncation of the number of final partial waves is necessary because no multipole decomposition is required and the contribution of all partial waves is included in a closed form. On the other hand, this approach does not appear as a systematic expansion of a perturbation series and thus, it is not obvious how to improve upon this approach in a systematic manner. In fact, in the applications higher-order terms are often included only if one fails to reproduce certain observables. Their selection appears more motivated by intuition than by a systematic procedure.

4.5 Covariant theory and dispersion relations

The first attempt towards a covariant description has been given by Pearlmann and Klein (PeK 60). Using field-theoretical methods they derived a formal expression for the S-matrix in terms of relativistic amplitudes for the bound and scattering states. This general S-matrix is then approximated by reducing the relativistic amplitudes to non-relativistic ones including only one-meson-exchange contributions. In this step covariance is lost. The meson-exchange-current contributions are related to the Born and resonance terms of the pion photoproduction amplitude. Explicit expressions for the $E1$ and $M1$ amplitudes including meson-exchange and Δ-resonance contributions are derived and the validity of Siegert's hypothesis is demonstrated, i.e., that the charge density is not affected by meson exchange in the static limit.

Unfortunately, this potentially powerful method being very approximate in the initial explicit evaluation has not been further developed. For this reason we will not describe it in detail. More popular has been the covariant approach using dispersion relations which has been started by Sakita and Goebel (SaG 62) and then in much greater detail developed by Le Bellac et al. (LeR 64 a,b). Later the Δ-excitation has been incorporated by George (Geo 68), Hasselmann (Has 68) and Wynn (Wyn 71), and more recently the question of possible dibaryon resonances has been studied within this approach by Huneke (Hun 80), Ogawa et al. (OgK 80) and Kang (Kan 86). For completeness, we would like to mention a different parametrization of the current matrix elements in terms of invariant form factors by Smirnov and Trubnikov (SmT 77).

The starting point of the covariant approach of SaG 62 is the observation that the twelve independent amplitudes of the T-matrix can be expressed in terms of twelve independent invariant operator amplitudes I_α in the following way

$$T(p_1,p_2,k,d) = \bar{u}(p_1)\sum_\alpha H_\alpha(s,t,u)I_\alpha C\bar{u}^T(p_2)\,, \tag{4.5.1}$$

where $C = i\gamma^2\gamma^0$ is the charge-conjugation operator of a Dirac particle. We use the metric and the definition of γ-matrices of BjD 64. The invariant scalar functions $H_\alpha(s,t,u)$ depend on the Mandelstam variables

$$s = (p_1+p_2)^2 = (k+d)^2 \tag{4.5.2}$$

$$t = (k-p_1)^2 \tag{4.5.3}$$

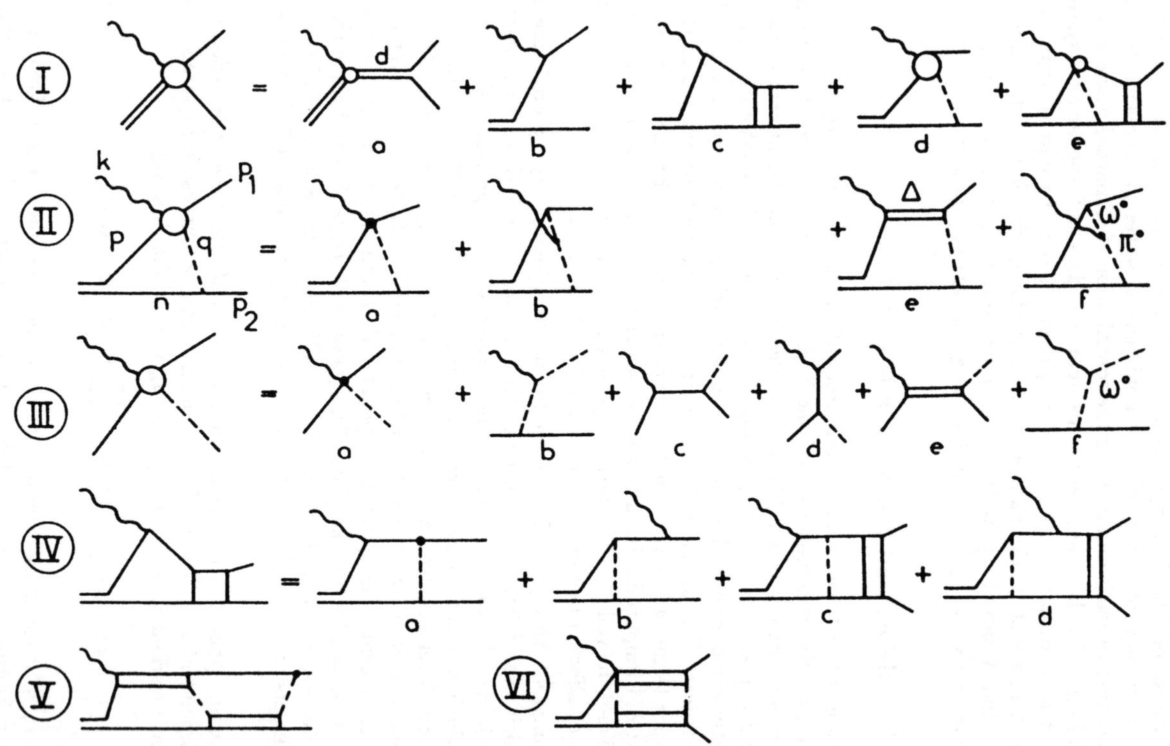

Fig. 4.4.1: Diagrams considered by Laget (Lag 78): (Ia) deuteron pole, (Ib) nucleon pole, (Ic) *N-N* final-state interaction with decomposition in (IV), (Id) pion photoproduction and reabsorption with decomposition in (II) and pion photoproduction part in (III). Two-loop diagrams in (V) and (VI).

$$u = (k - p_2)^2 \tag{4.5.4}$$

which fulfill the relation

$$s + t + u = 2M^2 + M_d^2 \,. \tag{4.5.5}$$

The invariant operators are of the general form

$$I_\alpha = \varepsilon_\mu I_\alpha^{\mu\nu} U_\nu \,, \tag{4.5.6}$$

where ε_μ and U_ν denote the polarization four-vectors of photon and deuteron, respectively. The $I_\alpha^{\mu\nu}$ are both Lorentz tensors and Dirac matrices and are constructed from the 16 Dirac matrices $1, \gamma_\mu, [\gamma_\mu, \gamma_\nu], \gamma_5\gamma_\mu, \gamma_5$ and the available independent four-vectors, i.e., the photon momentum k,

$$Q = \frac{1}{2}(p_1 + p_2) \tag{4.5.7}$$

and

$$q = \frac{1}{2}(p_1 - p_2) \,, \tag{4.5.8}$$

where p_i denote the outgoing nucleon momenta. Instead of p and P as defined in (3.1.13-14) we use in this section throughout q and Q, because it conforms with the standard notation of the covariant approach. For the same reason the photon momentum is denoted by k instead of (3.1.3).

The invariant operators should not contain $\not{q}$ and $\not{Q}$ because they are evaluated between the Dirac spinors in (4.5.1). They are constrained by the gauge-invariance condition

$$k_\mu I_\alpha^{\mu\nu} = 0 \,. \tag{4.5.9}$$

One finds first 14 invariants N_α as listed in table 4.5.1 which obey the gauge condition, where as a shorthand we have introduced for any two four-vectors a and b

$$N(a, b) = a_\mu(\varepsilon^\mu k^\nu - k^\mu \varepsilon^\nu) b_\nu \tag{4.5.10}$$

and

$$\begin{aligned} I(a) &= i\gamma_5 \epsilon^{\mu\nu\rho\sigma} U_\mu \varepsilon_\nu k_\rho a_\sigma \\ &= N_{14}\not{a} + \frac{1}{2}(N(\gamma,\gamma)U \cdot a + [N(a,\gamma), \not{U}]) \,. \end{aligned} \tag{4.5.11}$$

Possible gauge-invariant operators $I(q)$ and $I(Q)$ do not appear because these operators taken between the Dirac spinors are equivalent to linear combinations of other invariants according to (4.5.1), i.e.,

$$I(q) \hat{=} -\frac{1}{4}N_{10} - \frac{1}{2}N_{13} \tag{4.5.12}$$

$$I(Q) \hat{=} -MN_{14} - \frac{1}{2}N_{11} - \frac{1}{2}N_{12} \,. \tag{4.5.13}$$

The equivalence sign "$\hat{=}$" means that the operators on both sides of an equivalence relation give the same result if evaluated between the Dirac spinors in (4.5.1). But the equivalence cannot be used in operator products.

Tab. 4.5.1 Gauge-invariant amplitudes N_α

α	N_α	α	N_α	α	N_α
1	$N(U,q)$	6	$N(Q,q)\not{U}$	11	$N(\gamma,\gamma)U\cdot q$
2	$N(U,Q)$	7	$N(U,\gamma)$	12	$[\not{U},N(\gamma,q)]$
3	$N(Q,q)U\cdot q$	8	$N(\gamma,Q)U\cdot k$	13	$[\not{U},N(\gamma,Q)]$
4	$N(\gamma,Q)U\cdot q$	9	$N(\gamma,q)U\cdot k$	14	$-I(\gamma)$
5	$N(\gamma,q)U\cdot q$	10	$N(\gamma,\gamma)U\cdot k$		

These 14 invariants are further reduced to 12 by two additional relations which have first been derived by Sakita. Since we have not found their derivation in the literature, we will give one in the appendix D. Thus one can eliminate two of the 14 operators N_α and obtains finally 12 independent invariant operators corresponding to the fact that this process is described by 12 independent amplitudes as discussed in section 3.2. A standard set of these independent invariant operators is listed in table 4.5.2 taken from LeBellac et al. (LeR 64a). A different set of invariants had been introduced by Donnachie (Don 62b) but, as has been discussed by O'Donnell (ODo 63), the choice of SaG 62 is superior being free of kinematical singularities.

Tab. 4.5.2 Standard set of invariant operator amplitudes I_α

α	I_α	$\epsilon(\alpha)$	α	I_α	$\epsilon(\alpha)$
1	$\frac{1}{2M^2}N(q,U)$	$+$	7	$\frac{1}{2M^3}(N(q,\gamma)U\cdot k+2N(Q,\gamma)U\cdot q)$	$-$
2	$\frac{1}{2M^2}N(U,Q)$	$-$	8	$\frac{1}{4M^2}N(\gamma,\gamma)U\cdot k$	$+$
3	$\frac{1}{M^4}N(Q,q)U\cdot q$	$-$	9	$\frac{1}{2M^2}N(\gamma,\gamma)U\cdot q$	$-$
4	$\frac{1}{M^3}N(\gamma,Q)U\cdot q$	$-$	10	$\frac{1}{2M^2}([\not{U},N(\gamma,q)]+2N(Q,U))$	$-$
5	$\frac{1}{2M}N(U,\gamma)$	$+$	11	$\frac{1}{2M^2}([\not{U},N(\gamma,Q)]+2N(q,U))$	$+$
6	$\frac{1}{2M^3}(N(Q,\gamma)U\cdot k+2N(\gamma,q)U\cdot q)$	$+$	12	$\frac{1}{2M}I(\gamma)$	$-$

Up to now, the isospin of the final p-n state has not been considered. The total antisymmetry under interchange of the final two nucleons allows to separate the isospin components ($I=0$ and 1). The interchange $p_1\leftrightarrow p_2$, i.e., $q\rightarrow -q$ and $t\leftrightarrow u$ leads to

$$\begin{aligned} T(p_2,p_1,k,d) &= \bar{u}(p_2)\sum_\alpha H_\alpha(s,u,t)I_\alpha(-q)C\bar{u}^T(p_1) \\ &= \bar{u}(p_1)\sum_\alpha \epsilon(\alpha)H_\alpha(s,u,t)I_\alpha(q)C\bar{u}^T(p_2)\,, \end{aligned} \tag{4.5.14}$$

where the property

$$(I_\alpha(-q)C)^T = \epsilon(\alpha)I_\alpha(q)C \tag{4.5.15}$$

has been used. The sign function $\epsilon(\alpha)$ is also listed in table 4.5.2. Then one finds for the isospin amplitudes ($I = 0$ and 1)

$$T^I(p_1, p_2, k, d) = \frac{1}{2}\bar{u}(p_1)\sum_{\alpha}(H_\alpha(s,t,u) + (-)^I \epsilon(\alpha) H_\alpha(s,u,t) I_\alpha C \bar{u}^T(p_2) . \quad (4.5.16)$$

The helicity amplitudes can then be expressed in terms of the invariant functions $H_\alpha(s,t,u)$ (LeR 64a). Thus the main task in this approach is their determination which are assumed to possess a Mandelstam representation (MaV 61). Their analytic structure is obtained by investigating the singularities of the various helicity amplitudes as obtained from the different Feynman diagrams contributing to this process. A set of lowest-order diagrams usually considered is shown in fig. 4.5.1.

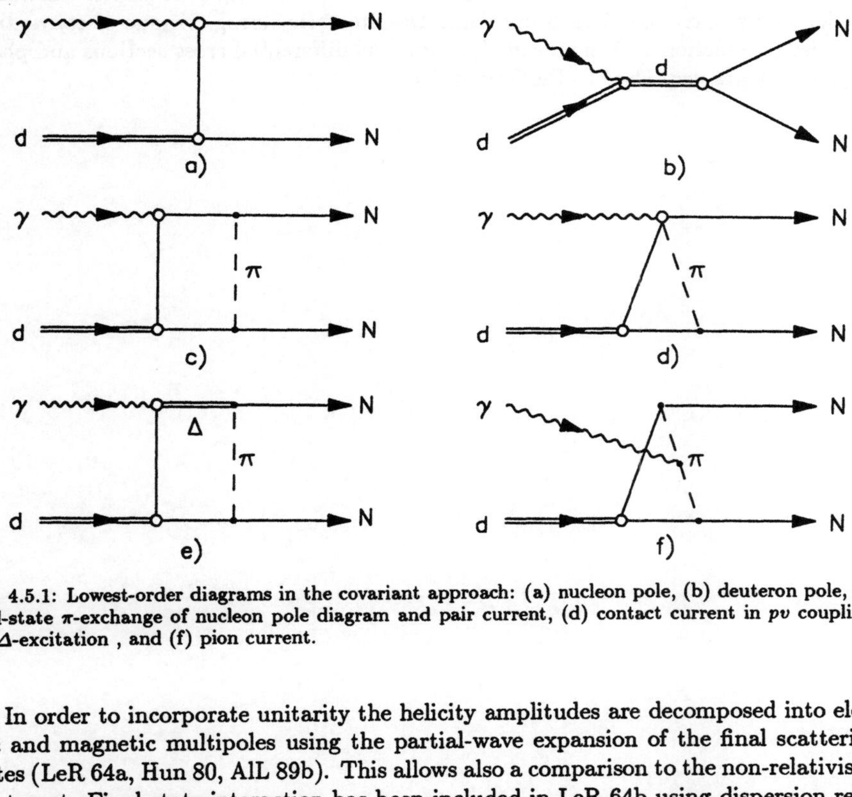

Fig. 4.5.1: Lowest-order diagrams in the covariant approach: (a) nucleon pole, (b) deuteron pole, (c) final-state π-exchange of nucleon pole diagram and pair current, (d) contact current in pv coupling, (e) Δ-excitation , and (f) pion current.

In order to incorporate unitarity the helicity amplitudes are decomposed into electric and magnetic multipoles using the partial-wave expansion of the final scattering states (LeR 64a, Hun 80, AlL 89b). This allows also a comparison to the non-relativistic treatment. Final-state interaction has been included in LeR 64b using dispersion relations with N-N phase shifts from experiment as input. It is found that the agreement with experimental data for total and differential cross sections, for photon asymmetry and outcoming-nucleon polarization from threshold to $E_\gamma = 130$ MeV is as good as the non-relativistic description.

The low-energy regime has received special attention. Sakita (Sak 62) has derived a low-energy limit as a generalization of the Kroll-Ruderman theorem of pion photoproduction. The zero-energy limit of the transition amplitude with respect to the incident photon is expressed in terms of static properties of nucleon and deuteron, deuteron binding energy and triplet effective range of p-n scattering. This result has been used to derive the $E1$ and $M1$ contributions to total and differential cross sections and polarization observables in a model-independent way (GoL 81, LuM 85, AlL 89a). While the total cross section is well described up to $E_\gamma = 80$ MeV, angular distributions show larger deviations and the polarization observables cannot be described at all by the low-energy theorem (AlL 89a). Furthermore, meson-exchange current contributions to thermal n-p capture have been studied by Skolnick (Sko 64) using the covariant approach (see 7.1).

Finally, we would like to mention another covariant approach by Nagornyi et al. (NaK 86). They use a field-theoretical description of a compound object with internal structure. Its wave function is described in terms of light-front dynamics. In the actual numerical evaluation they rely on quite a few approximations in using realistic nonrelativistic potentials like Reid and Paris instead of deriving a consistent interaction model within their covariant framework. Despite their strong critique of conventional theories (see section 4.2) the results for total and differential cross sections and photon asymmetry are very close to the former ones.

5 Unpolarized and polarized beam sources

The cross sections of photon-induced nuclear processes were usually studied with the aid of a bremsstrahlung beam produced by fast electrons hitting a suitable radiator. Such bremsstrahlung radiation contains photons of all energies from zero up to the kinetic energy of the incoming electrons and so the desired cross section had to be deduced from the integral yield by taking the bremsstrahlung spectrum of the photon beam from the theory. In this section we shall show that this method can produce significant systematic errors in the absolute value of the cross sections due to the lack of a precise knowledge of the bremsstrahlung spectrum.

Then several other methods will be presented that have been applied in the last decades in order to obtain monochromatic photon beams or at least to measure directly the number of photons employed in the measurement. Moreover, we shall discuss the production of neutron beams of high energy and intensity, since they have been used in the study of the inverse *n-p* capture reaction. Finally we shall review the methods applied for the production of polarized photon and neutron beams and discuss the accuracy of the measured degree of polarization.

5.1 Continuous bremsstrahlung spectrum and the monitoring problem

Except for a few cases, the source of photons usually employed in photonuclear experiments has been a continuous bremsstrahlung beam. Since the process of interest here involves two particles only in the final state, the incident-photon energy has been inferred from the measured energy and/or angle of the emitted particles using kinematical relations.

The number of photons is deduced from the energy integrated beam power measured by a quantameter and from the distribution of photons as a function of energy, which is taken from the theory. Should the bremsstrahlung spectrum be different from what is known theoretically, the deduced number of photons for a fixed energy and therefore the measured absolute cross section would require a readjustment. As a consequence an accurate knowledge of the bremsstrahlung spectrum is essential for measurements of absolute cross sections.

The calculation of the radiation by electrons in the field of a nucleus and of the atomic electrons is rather complicated. The problem of evaluating an "exact" expression for the bremsstrahlung cross section for the electron-nuclear interaction involves for the matrix element the use of "exact" wave functions, which describe an electron in a screened nuclear Coulomb field. Cross section calculations for electron-electron bremsstrahlung are furthermore complicated because of the exchange character of the interaction due to the Pauli principle. Up to now no single analytic expression is available, which is valid for all electron energies or all radiators or even for the entire photon spectrum for a given electron energy and radiator.

An extensive collection of bremsstrahlung formulae with detailed discussions of the assumptions and approximations can be found in the old review of Koch and Motz (KoM

59). Among all the calculations reported in KoM 59 the formula of Schiff and the one of Olsen and Maximon are selected for the following discussion. In the work of Schiff (Sch 51) relatively simple analytic formulae have been given by the use of "no screening" – or "complete screening" – conditions for the electron in the nuclear Coulomb field and by including the influence of electron-electron bremsstrahlung replacing the Z^2-dependence of the cross section by $Z(Z+1)$. Olsen and Maximon (OlM 59) include screening corrections using the Thomas-Fermi-Molière form factor for the atomic electrons. Also in this work only the bremsstrahlung in the field of the nucleus is evaluated.

In the case of "no screening" or "complete screening" conditions the calculations of OlM 59 and the one of Sch 51 give similar results apart from Coulomb corrections introduced by Olsen and Maximon. This Coulomb correction, even if it affects the cross section only by a few percent, generates an appreciable distortion in the photon spectrum depending on the atomic number of the radiator.

In fig. 5.1.1 the relative intensity $N(E_\gamma) \cdot E_\gamma$ is shown for 40 MeV and 300 MeV electron energy as a function of the photon energy for several photon emission angles calculated in KoM 59 using the formula of Schiff and the formula of Olsen and Maximon (with correct screening). In fig. 5.1.2 we show the relative bremsstrahlung intensities integrated over the photon emission angle as calculated in KoM 59 using various formulae, again for 40 MeV and 300 MeV electron energy. It must be noted that the cross section integrated over the photon emission angle of OlM 59 and the one of Bethe and Heitler (BeH 34) are identical apart from the Coulomb correction factor of OlM 59.

From the comparison of fig. 5.1.1 and fig. 5.1.2 it becomes evident that a large discrepancy exists in the shape of the differential cross section with respect to the photon angle evaluated using different formulae, while different calculations produce only a little difference in the shape of the cross section integrated over the photon angle. Moreover, since the bremsstrahlung spectrum is very dependent on the photon emission angle, the shape of the photon spectrum is sensitive to the collimation of the utilized photon beam and to electron scattering effects in the radiator. This means that, when in an experiment the photon beam has no collimation or a very weak collimation, any realistic bremsstrahlung calculation gives useful results for the number of employed photons. On the contrary, when due to strong collimation it is necessary to know very precisely the angular distribution, different assumed formulae can give very different results and the difference depends on the experimental condition (collimation, alignment, radiator thickness, etc.).

The only extensive calculation one can find in the literature concerning electron-electron bremsstrahlung was performed by Tsai (Tsa 65, Tsa 74). The approximations introduced in these calculations make the results accurate only for $\gamma\vartheta \ll 1$, where $\gamma = E_e/m_0$ with electron energy E_e and electron mass m_0 and ϑ is the photon emission angle. In table 5.1.1 we give the number of photons normalized to the total energy of the beam and evaluated using different formulae for $E_\gamma = 100$ MeV and $Z = 74$ as an example (San 86). The disagreement between the results depends on the energy region in the bremsstrahlung spectrum: it is not more than 15 percent in the central region but it can reach a factor of 2 or more near the end-point of the spectrum.

Recently very accurate calculations of the bremsstrahlung cross section integrated over the photon angle, for electron energies from 1 keV to 1 GeV, have been performed using Hartree-Fock atomic form factors for the screening effect and including very consistently Coulomb corrections for the bremsstrahlung in the Coulomb field of the atomic nucleus and an accurate treatment of the electron- electron bremsstrahlung (SeB

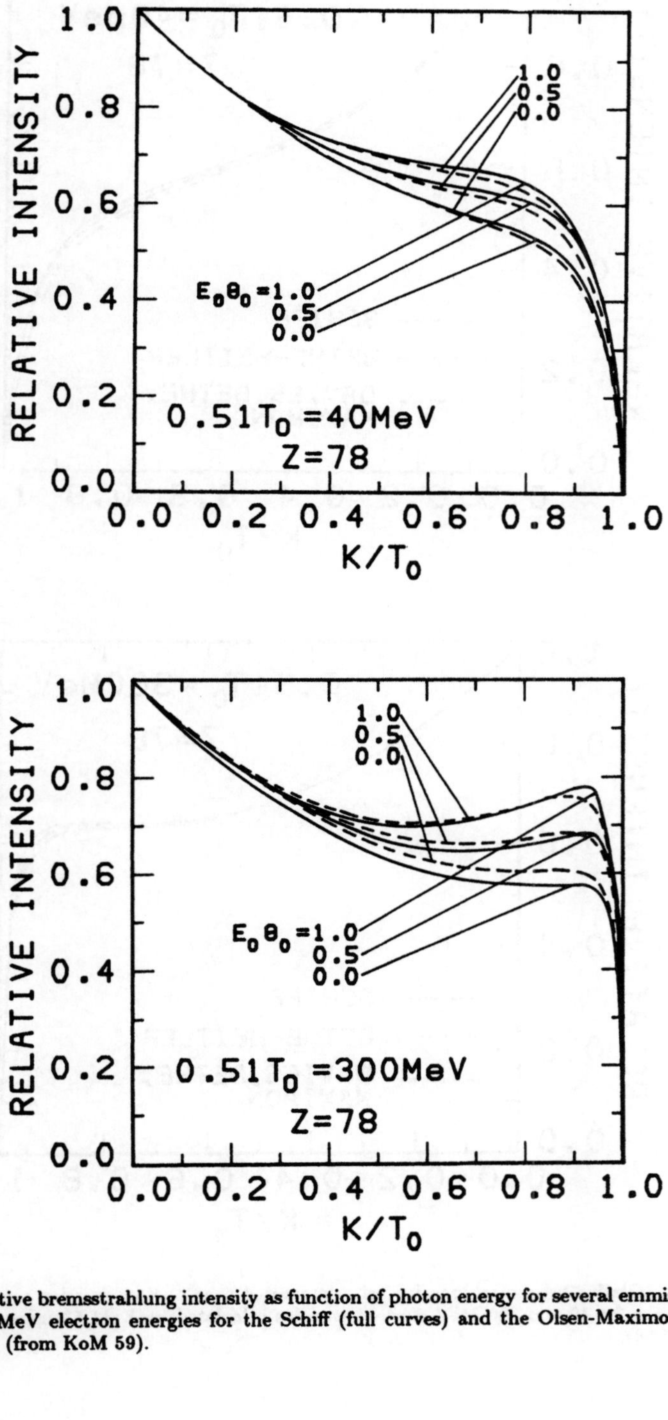

Fig. 5.1.1: Relative bremsstrahlung intensity as function of photon energy for several emmission angles at 40 and 300 MeV electron energies for the Schiff (full curves) and the Olsen-Maximon spectrum (dashed curves) (from KoM 59).

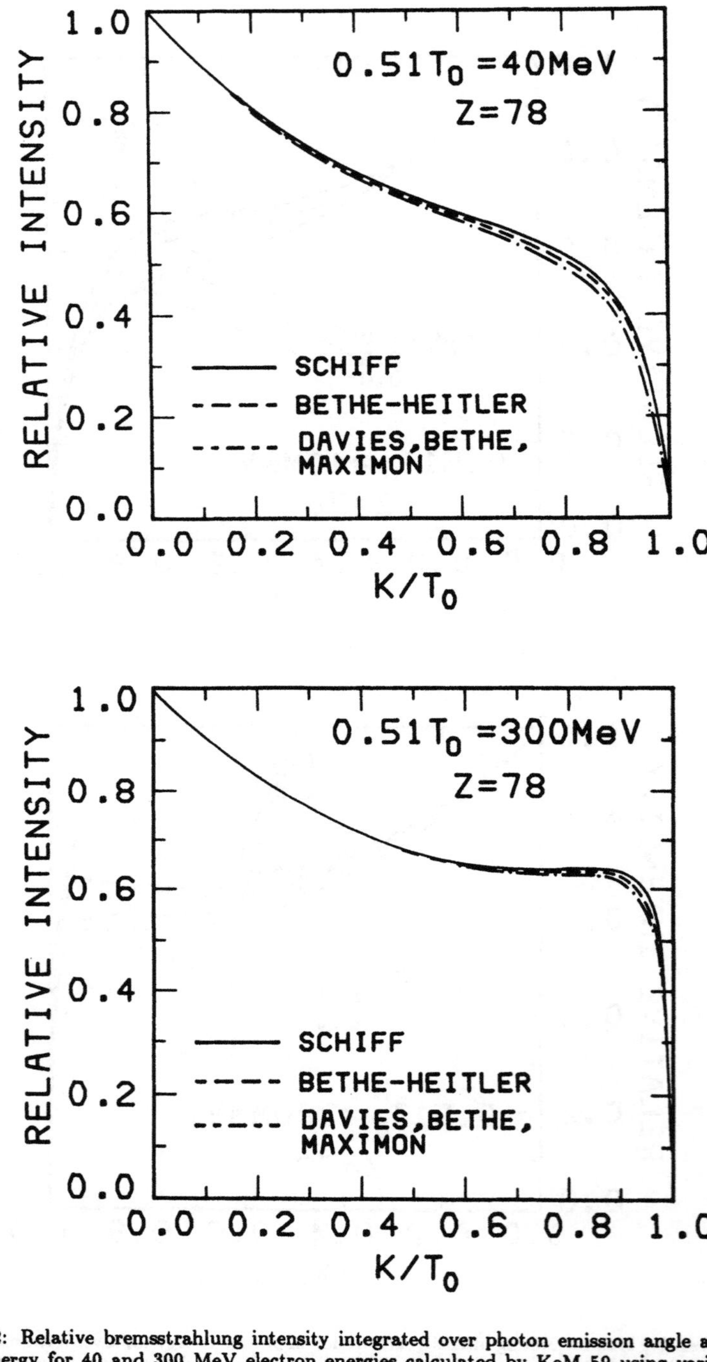

Fig. 5.1.2: Relative bremsstrahlung intensity integrated over photon emission angle as function of photon energy for 40 and 300 MeV electron energies calculated by KoM 59 using various formulae (from KoM 59).

Tab. 5.1.1 Number of photons normalized to the total energy (arbitrary units)

Formula (ref.)	Photon energy range $0.95E_{\gamma max}-E_{\gamma max}$	$0.90E_{\gamma max}-E_{\gamma max}$	$0.40E_{\gamma max}-0.60E_{\gamma max}$
Photon emission angle $\vartheta = 0$ rad			
Schiff without screening (Sch 51)	11.8	28.4	355
Schiff complete screening (Sch 51)	27.5	61.8	387
Olsen & Maximon (OlM 59)	26.0	58.7	389
Tsai (Tsa 74)	29.7	63.7	387
Photon emission angle $\vartheta = 8$ mrad			
Schiff without screening (Sch 51)	19.1	45.1	444
Schiff complete screening (Sch 51)	46.5	99.7	525
Olsen & Maximon (OlM 59)	27.9	64.4	450
Tsai (Tsa 74)	29.6	70.2	447

85, SeB 86). Unfortunately similar calculations concerning the differential cross section as a function of the photon emission angle are not available at present.

5.2 Monochromatic and quasi-monochromatic photon beams

In order to avoid the systematic errors arising from the use of a bremsstrahlung beam and from the monitoring procedure described in 5.1, two methods are available. The first one is the use of a bremsstrahlung beam with a direct measurement of the number of photons at a given energy. The second consists in the use of monochromatic or quasi-monochromatic γ-ray sources. A general overview of the most common experimental procedures can be found in BeB 80. Here we shall limit ourselves to the methods used in experiments on photodisintegration of the deuteron.

For what concerns the first method, a detailed knowledge of the photon spectrum can be obtained by using a magnetic spectrometer, i.e., a Compton or a pair spectrometer, on line during the measurement. In a Compton spectrometer the kinetic energy of the electrons emitted into a small cone in the forward direction is generally magnetically analyzed starting from

$$E_{\gamma'} = \frac{E_\gamma}{1+\frac{E_\gamma}{m_0}(1-\cos\Theta)}\,, \qquad (5.2.1)$$

where m_0 is the electron mass, $E_\gamma(E_{\gamma'})$ is the initial (final) energy of the photon, and Θ is the scattering angle of the photon. For $\Theta = \pi$ and $E_\gamma \gg m_0$ one has

$$E_{\gamma'} \simeq \frac{m_0}{2} \qquad (5.2.2)$$

$$E_e \simeq E_\gamma - \frac{m_0}{2}\,, \qquad (5.2.3)$$

where E_e is the energy of the recoiling electron emitted in this case at 0°.

Therefore, the measurement of the energy of the forward-emitted electron is a direct measurement of the photon energy. Moreover, for this kinematic condition the differential cross section is largest. The number of Compton electrons is determined knowing the efficiency of the spectrometer and subtracting the number of electrons coming from pair production. This is obtained by measuring separately the positrons from pair production with reversed magnetic fields. More details can be found in AhB 73.

In the pair spectrometer the photon energy is deduced from

$$E_\gamma = E_{e^-} + E_{e^+} + 2m_0\,. \qquad (5.2.4)$$

The kinetic energies of the electron and the positron emitted into a small cone in the forward direction are magnetically analyzed. Even in this case, in order to measure the number of photons in a given energy bin, the efficiency of the magnetic spectrometer must be known with a good accuracy (CaD 80). Both with Compton and pair spectrometers the intensity of photons can be measured with an accuracy of a few percent.

To determine the energy of the photon in a bremsstrahlung spectrum and the absolute number of photons in a given energy bin the tagging method can also be used. It consists in the measurement of the energy E' of the electron after having radiated a photon in coincidence with the reaction products induced by this photon. The decelerated electron is deflected by an analyzer magnet and detected by a counter hodoscope, thus tagging the presence and the energy $E_\gamma = E_0 - E'$ of the corresponding photon.

The scheme of a typical tagging system as used for deuteron photodisintegration is illustrated in fig. 5.2.1. In the tagging method the intensity of the beam is limited by accidental coincidences between the electron counters and the detector of the reaction products. Therefore, only with high-duty-cycle electron accelerators it is possible to utilize a high number of tagged photons. Unfortunately, high-duty-cycle accelerators have become available in the last decade only.

The number of photons of a given energy is obtained by counting the tagging signal without coincidence and by knowing accurately the efficiency of the tagging system. The latter is determined by placing a detector with well-known efficiency directly into the photon beam, for example, a Cerenkov counter of 10 radiation length thickness, as has been done in ArE 82 where details can be found. The advantage of a tagged photon

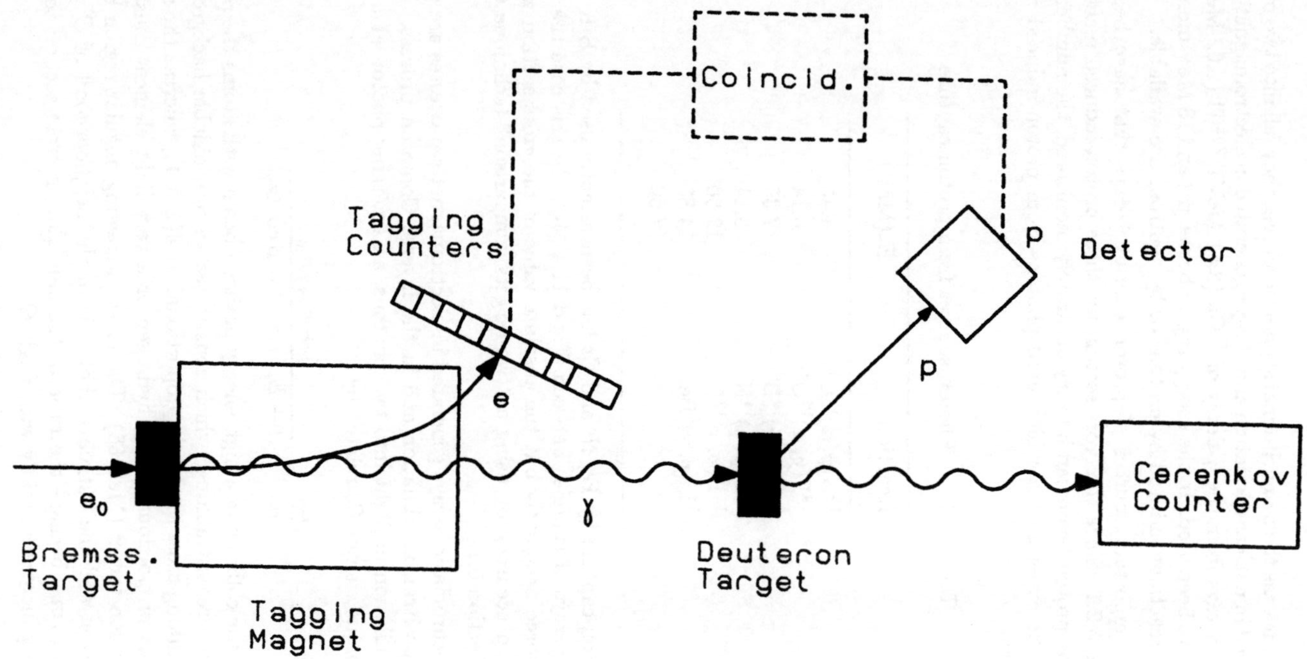

Fig. 5.2.1: Scheme of a typical tagging system as use for deuteron photodisintegration experiments.

beam is that it allows an accurate measurement of the absolute cross section, because no knowledge of the total beam power or the shape of the photon energy spectrum is required.

For what concerns the monochromatic photon sources they fall into two main classes, γ-rays produced in nuclear excitation and γ-rays generated in electromagnetic processes. Below 3 MeV γ-rays from the γ-decay of ^{72}Ga (2.51 MeV), ^{208}Tl (2.62 MeV) or ^{24}Na (2.76 MeV) have been used. In the energy region between 4 and 18 MeV monochromatic photons from reactions induced by protons or by neutrons are available.

The more important sources of proton capture γ-rays that have been used are listed in table 5.2.1. Since the cross section for these proton-induced reactions is well known and the proton beam can be very accurately monitored, the number of photons employed in the experiments, which used photons from proton-induced reactions, is precisely known.

Tab. 5.2.1 Photon sources from proton capture

Reaction	E_γ(MeV)
$^{15}N(p,\alpha)^{12}C^*$	4.45
$^{19}F(p,\alpha)^{16}O^*$	6.14
$^{9}Be(p,\gamma)^{10}B$	7.39
$^{13}C(p,\gamma)^{14}N$	8.14
$^{11}B(p,\gamma)^{12}C$	12.50
$^{7}Li(p,\gamma)^{8}Be$	14.80
	17.60

The neutron-capture by Fe, Ni and Cr has been recently used by BiK 85 and BiB 88 to obtain γ-rays of energies between 6 and 11.4 MeV. In this case the beam could not be monitored accurately but the absolute value of the cross section was deduced with very good accuracy by using a "two-targets comparison"-technique that will be discussed in section 6.1.

The monochromatic γ-rays generated in electromagnetic processes are produced in the electron-positron annihilation and in the inverse Compton process. In electron-positron annihilation in flight into two photons, an emerging photon at a fixed angle Θ_γ has an unique energy E_γ related by

$$\frac{E_\gamma}{m_0} = \frac{1 + E_{e^+}/m_0}{1 + E_{e^+}/m_0(1 - \beta\cos\Theta_\gamma)} \,. \tag{5.2.5}$$

Obviously, in the collision of a high-energy positron beam with atoms the predominant reaction is the bremsstrahlung, which constitutes an unavoidable background. Since the bremsstrahlung cross section is proportional to $Z(Z+1)$, whereas the annihilation cross section is proportional to Z, hydrogen or a very light element must be chosen as an annihilation target (MaS 68). The most interesting angular region between the incident positron and the outgoing photon is in the neighborhood of $\Theta = (2/\gamma)^{1/2}$, where $\gamma = E_{e^+}/m_0$, because the ratio of the annihilation cross section to the one of bremsstrahlung increases rapidly with angle Θ.

A typical photon spectrum obtained by CaD 83, when a positron beam impinges on a hydrogen target, is shown in fig. 5.2.2. To subtract from the total spectrum the continuous background due to bremsstrahlung two methods are available. The first consists in using positron and electron beams with identical characteristics and the same target. The second one is to use only a positron beam and a low-Z and a high-Z target of the same thickness in radiation-length, so that the bremsstrahlung contribution is the same, while the annihilation peak is negligible in the high-Z target.

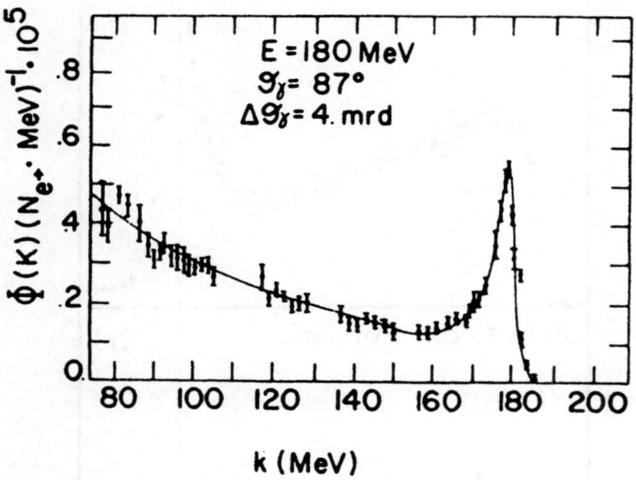

Fig. 5.2.2: Photon spectrum from 180 MeV positron annihilation in a liquid hydrogen target for photon collection angle $\Theta_\gamma = 0.87°$ (from CaD 83).

The results of the two methods are shown in figs. 5.2.3 and 5.2.4. It must be noted that due to screening effects and Coulomb corrections the bremsstrahlung spectra for a low-Z and a high-Z target with the same radiation-length thickness are not exactly the same, as discussed in section 5.1. Therefore in the subtraction a correction must be applied.

Monochromatic photons can also be obtained by Compton scattering of laser light off high-energy electrons. The process is presented in fig. 5.2.5. The energy E_{γ_2} of the scattered photon is given by

$$E_{\gamma_2} = E_{\gamma_1} \frac{1 - \beta \cos \Theta_1}{1 - \beta \cos \Theta_2 + \frac{E_{\gamma_1}}{E_e}(1 - \cos \Theta)} , \tag{5.2.6}$$

with the notation:

E_{γ_1} energy of the incident photon,
E_e energy of the incident electron,
E_{γ_2} energy of the scattered photon,
β electron velocity in units of c,
Θ_1 angle between the direction of the incident electron and the incident photon,
Θ_2 angle between the direction of the incident electron and the scattered photon.

For $\Theta_1 = \pi, \beta \simeq 1$ and $\Theta_2 = 0$ one finds

$$E_{\gamma_2 max.} = \frac{2E_{\gamma_1}}{1 - \beta + 2\frac{E_{\gamma_1}}{E_e}} . \tag{5.2.7}$$

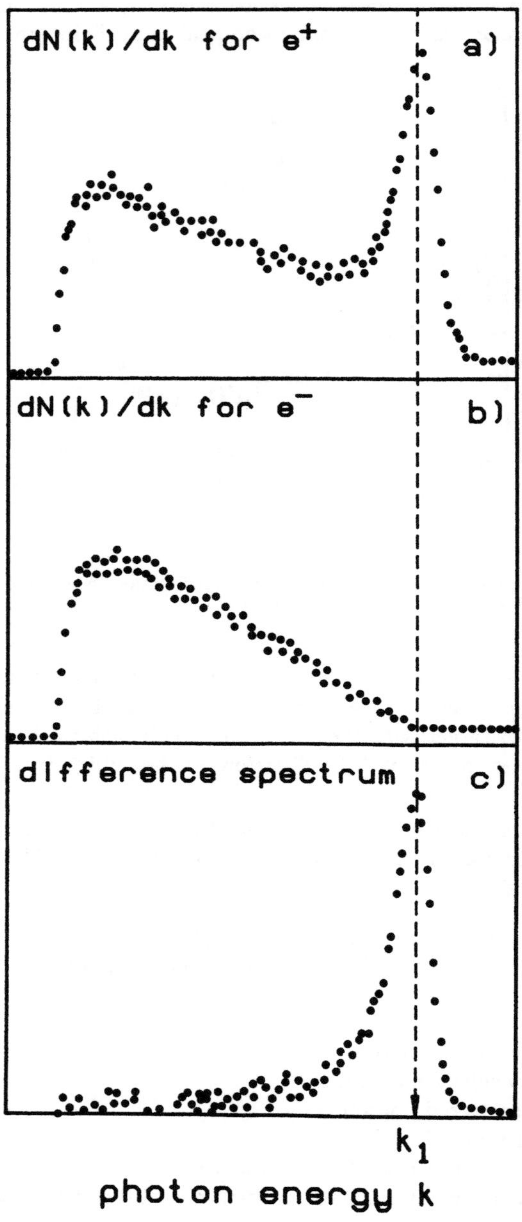

Fig. 5.2.3: Measured photon spectra obtained with identical e^+ (a) and e^- (b) beam conditions and the difference spectrum (c) (from BeB 80).

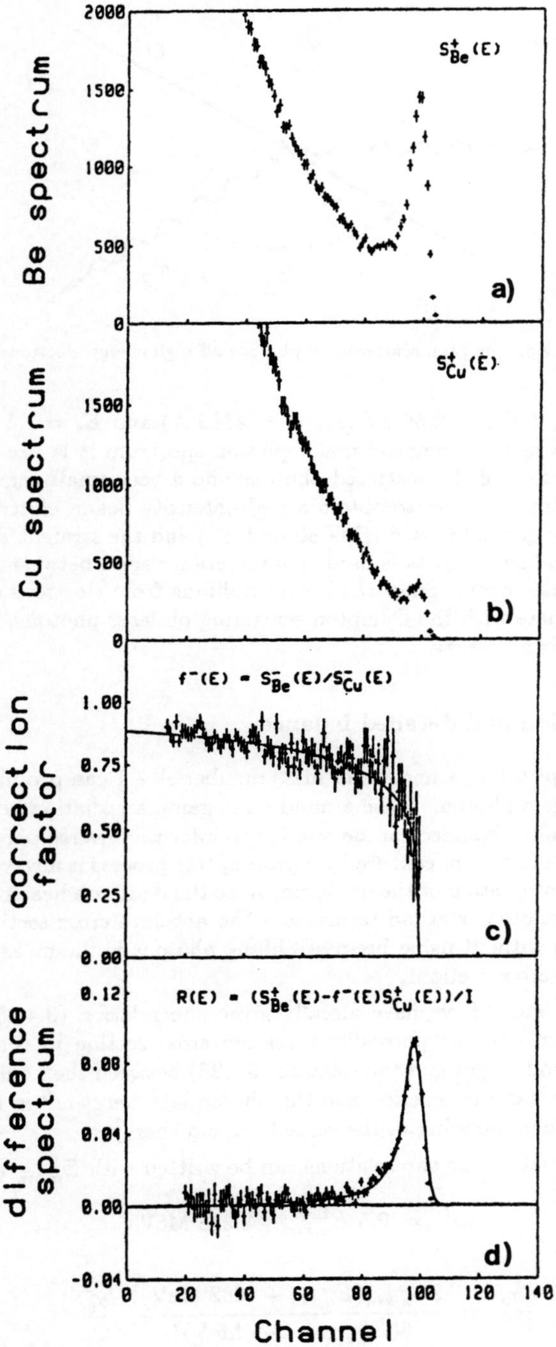

Fig. 5.2.4: Photon spectra obtained with a positron beam and Be (a) and Cu (b) radiators and the difference spectrum (d) corrected for the difference between low- and high-Z bremsstrahlung spectra by the factor $f^{-}(E)$ (c) (from LeS 81).

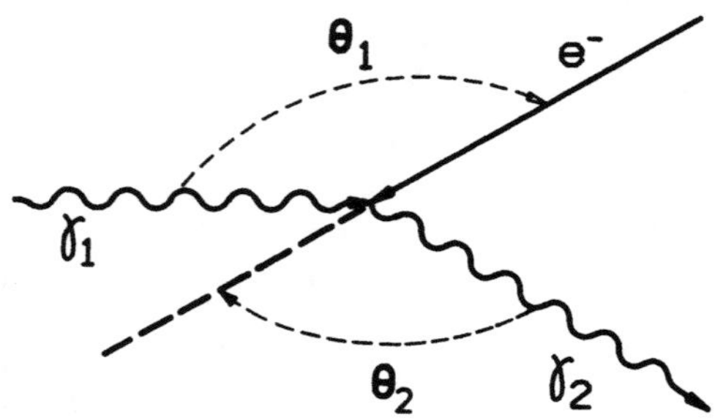

Fig. 5.2.5: Compton scattering of photons off high-energy electrons.

As an example, if $E_{\gamma_1} = 2.56$ eV (i.e. $\lambda = 4880$ Å) and $E_e = 1.5$ GeV, one gets $E_{\gamma_2} = 83$ MeV. To have a monochromatic photon spectrum it is necessary to use a very narrow collimation of the scattered photons and a very small angular divergence for the electron beam. In order to obtain a high-intensity beam, electrons circulating in a storage ring are generally used (FeG 80, SaL 83) and the straight section between two consecutive bending magnets is used as interaction region between the circulating electrons and the laser beam. Since the bremsstrahlung from electrons on residual gas molecules will compete with the Compton scattering on laser photons, a high vacuum in the storage ring is necessary.

5.3 Inverse reaction and detailed balance

The nucleon capture by a nucleus of mass number $A-1$ can produce a nucleus of mass number A plus a photon. If the ground-state gamma radiation is observed, then the principle of detailed balance can be applied to infer the inverse photonuclear cross section. In the case of neutron capture by a proton, this process is always related to the two-body photodisintegration of the deuteron, since the deuteron has no bound excited states. Therefore, another method to measure the absolute cross section of deuteron photodisintegration without using bremsstrahlung photons is the measurement of the inverse n-p capture cross section.

In sections 3.1 and 3.4 we have already given the relation (3.4.25) between the capture cross section and the photodisintegration cross section in the cm frame for unpolarized beam and target and the relation (3.1.26) between the neutron lab kinetic energy $E^L_{n,kin}$ in the capture reaction and the photon lab energy ω^L in the photodisintegration process, corresponding to the same total cm energy.

In first approximation the two relations can be written with $E^L_{n,kin}$ and ω^L in MeV

$$\omega^L = 0.5\, E^L_{n,kin} + 2.225 \text{ MeV} \tag{5.3.1}$$

and

$$\frac{d\sigma_0}{d\Omega_p} = \frac{2E^L_{n,kin}(E^L_{n,kin} + 1882 \text{ MeV})}{3(E^L_{n,kin} + 4.45 \text{ MeV})^2} \frac{d\sigma_0^c}{d\Omega_\gamma} . \tag{5.3.2}$$

Obviously the two differential cross sections refer to the same cm angle. In order to find the relation between the proton emission lab angle in $\gamma + d \to n + p$ and the photon emission lab angle in $n + p \to d + \gamma$ corresponding to the same cm angle, we give the kinematical relation between lab and cm frame angles for the two reactions. Denoting by Θ the angle in the cm system and in the lab frame by Θ_p^L the angle of the emitted proton in the photodisintegration reaction and by Θ_γ^L the angle of the emitted photon in the capture process, one gets in the approximation $M_p = M_n = 1/2\ M_d = M$ for the $d(\gamma, p)n$ reaction

$$\cot \Theta_p^L = \gamma_p \frac{\cos \Theta + \rho_p}{\sin \Theta} \tag{5.3.3}$$

$$\cos \Theta = \frac{-\gamma_p^2 \rho_p \pm \cot \Theta_p^L \sqrt{\gamma_p^2(1-\rho_p^2) + \cot^2 \Theta_p^L}}{\gamma_p^2 + \cot^2 \Theta_p^L} \tag{5.3.4}$$

$$\sin \Theta = \frac{\gamma_p \rho_p \ \cot \Theta_p^L \pm \sqrt{\gamma_p^2(1-\rho_p^2) + \cot^2 \Theta_p^L}}{\gamma_p^2 + \cot^2 \Theta_p^L} \tag{5.3.5}$$

with

$$\gamma_p = \frac{2+\epsilon_\gamma}{2\sqrt{1+\epsilon_\gamma}} \tag{5.3.6}$$

$$\rho_p = \frac{\sqrt{\epsilon_\gamma(1+\epsilon_\gamma)}}{2+\epsilon_\gamma}, \tag{5.3.7}$$

where

$$\epsilon_\gamma = \frac{\omega^L}{M}, \tag{5.3.8}$$

and for the $p(n, \gamma)d$ reaction

$$\cot \Theta_\gamma^L = \gamma_\gamma \frac{\cos \Theta + \rho_\gamma}{\sin \Theta} \tag{5.3.9}$$

$$\cos \Theta = \frac{-\gamma_\gamma^2 \rho_\gamma \pm \cot \Theta_\gamma^L \sqrt{\gamma_\gamma^2(1+\rho_\gamma^2) + \cot^2 \Theta_\gamma^L}}{\gamma_\gamma^2 + \cot^2 \Theta_\gamma^L} \tag{5.3.10}$$

$$\sin \Theta = \frac{\gamma_\gamma \rho_\gamma \ \cot \Theta_\gamma^L \pm \sqrt{\gamma_\gamma^2(1+\rho_\gamma^2) + \cot^2 \Theta_\gamma^L}}{\gamma_\gamma^2 + \cot^2 \Theta_\gamma^L} \tag{5.3.11}$$

with

$$\gamma_\gamma = \sqrt{\frac{\epsilon_n + 2}{2}} \tag{5.3.12}$$

$$\rho_\gamma = \sqrt{\frac{\epsilon_n}{\epsilon_n + 2}}, \tag{5.3.13}$$

where

$$\epsilon_n = \frac{E_{n,kin}^L}{M}. \tag{5.3.14}$$

The neutron beam is generally produced by a (p, n) reaction induced in a low-Z target by protons accelerated in a cyclotron. The most common process used for this has been ^{7}Li(p, n). But also reactions like ^{9}Be(p, n), ^{9}Be(d, n), ^{3}H(d, n) and ^{2}H(p, n) have been employed.

In principle the monitoring of the neutron intensity is difficult and the measurement cannot be done with high precision. For this reason the cross section for the capture reaction has often been deduced by measuring at the same time deuterons and protons from the reactions $p(n,d)\gamma$ and $p(n,p)n$ and assuming as well-known the differential cross section for the $p(n,p)n$ process. The neutron energy has been measured with time-of-flight techniques.

It must be noted, however, that the data concerning the elastic scattering cross sections of $p(n,p)n$ show strong fluctuations and that the theoretical predictions are in substantial disagreement with the data, as shown in fig. 5.3.1 as an example, concerning the neutron energy range $E^L_{n,kin} = 180 - 200$ MeV.

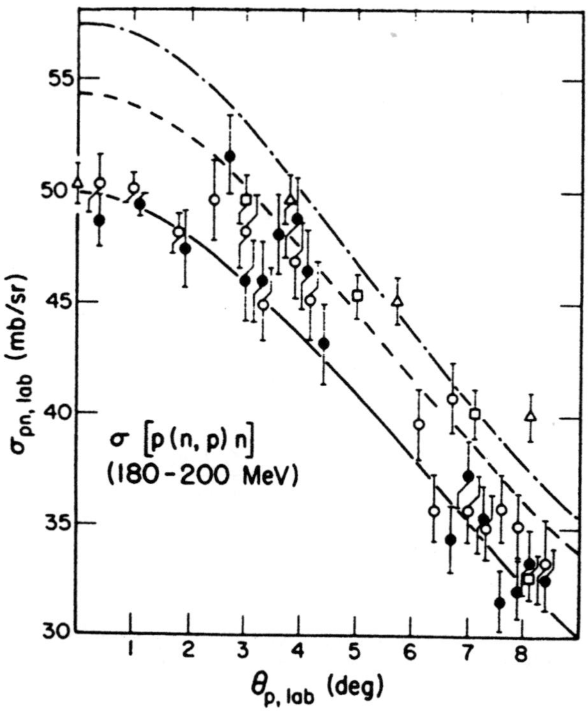

Fig. 5.3.1: Elastic lab cross section for $p(n,p)n$ at $E^L_{n,kin} = 180-200$ MeV. Comparison of experimental data with different calculations (from MeH 85).

As a consequence the absolute value of the cross section for the capture process, evaluated with this method, cannot have a precision better than 10 percent in general. Only in one case (DuL 85), the protons scattered at 90° in the cm system were used for the normalization. Since in the energy range used in this experiment the 90°-cross section can be deduced with an error of the order of 1.5 percent, the precision of the capture cross section is better than in other cases.

Before concluding this section we want to make two comments concerning the n-p capture experiments. The first one is that a neutron beam contains always a high γ-ray contamination and only a good time-of-flight technique for measuring the neutron energy can reduce drastically this contamination. The second comment is that at energies above pion production threshold the deuteron spectrum, even if measured in coincidence with the emitted photon, can have contributions from the competing process $n+p \to d+\pi^{\circ}(\to 2\gamma)$ which is kinematically very similar to the simple capture reaction.

5.4 Polarized beam sources

In the following we shall discuss the experimental methods and techniques used for the production of polarized beams. While monochromatic beams improve the accuracy of the photoreaction experiments, the use of polarized beams is expected to give further insight into the dynamical properties of the system under study. Therefore, in the last decades several methods for obtaining both polarized photon beams and polarized neutron beams have been developed and applied successfully. We shall review the methods employed in experiments concerning the photodisintegration of the deuteron and its inverse reaction.

Polarized photons have been obtained from

i) angular selection of γ-rays emitted in a capture reaction like ${}^3\mathrm{H}\ (p,\gamma)\ {}^4\mathrm{He}$,
ii) angular selection of bremsstrahlung from very thin targets,
iii) bremsstrahlung in crystals without any photon angular selection,
iv) Compton backscattering of laser photons by high-energy electrons.

We shall now discuss these methods.

i) Let us consider the reaction ${}^3\mathrm{H}\ (p,\gamma)\ {}^4\mathrm{He}$. Due to the high Q-value ($Q = 19.81$ MeV) this reaction can produce γ-rays of relatively high energy. As an example, for an incident proton of energy of 1 MeV the average photon energy is $E_\gamma \approx 20.3$ MeV. Moreover, the γ-rays emitted at an angle $\Theta_\gamma = 90°$ have a high degree of linear polarization. In fact, the polarization of the γ-rays depends on the relative strength of the allowed multipole transitions and hence on their angular distribution. At small proton energies the reaction is expected to proceed mainly through $E1$, $M1$, and $E2$ transitions with $E1$ being dominant. As a consequence the differential cross section can be written in the form

$$\sigma(\Theta_\gamma) = A + B\sin^2\Theta_\gamma + C\sin^2\Theta_\gamma\cos\Theta_\gamma + D\sin^2\Theta_\gamma\cos^2\Theta_\gamma\ . \tag{5.4.1}$$

In table 5.4.1 taken from DBK 80 the angular distribution and the associated polarization for these multipole transitions are shown. At the angle $\Theta_\gamma = 90°$ the γ-ray flux is proportional to the quantity $A+B$: the γ-rays associated with the term B result from ordinary ($\Delta S = 0$) $E1$ transitions and are linearly polarized with the electric vector parallel to the reaction plane. Since at $E_p \approx 1$ MeV the ratio A/B is of the order of 10^{-3} (DBK 80), it follows that the degree of polarization for the γ-rays emitted at $\Theta_\gamma = 90°$ is very close to unity.

ii) This method is very simple in principle, but encounters rather large experimental difficulties. The high-energy bremsstrahlung photons are linearly polarized in a direction perpendicular to their emission plane. This polarization has a complete axial symmetry and a polar-angle dependence that reaches a maximum at the angle m_0/E_e where m_0 denotes the electron mass, while it must be zero in the forward direction for symmetry reasons. Therefore, the whole bremsstrahlung beam is almost completely unpolarized, but a polarized beam of photons may be obtained by selecting a small solid angle around the angle m_0/E_e. The expected polarization is a decreasing function of the ratio $x = E_\gamma/E_e$. Typical theoretical values are $P = 0.50$ for $x = 0.25$ and $P = 0.30$ for $x = 0.63$. Of course, the main problem is the competition between the multiple scattering in the radiator and the size and angular divergence of the electron beam with respect to the angle m_0/E_e. This competition causes a decrease of the polarization and therefore must be minimized. With this method very thin radiator targets must then be used and the angular selection is obtained by placing an appropriate off-axis collimator between the radiator and the experimental target. As a consequence the photon intensity is drastically reduced in comparison with a normal bremsstrahlung beam. The degree of polarization that one can obtain with this method is of the order of $0.3-0.6$.

Tab. 5.4.1 $^3H(p,\gamma)^4He$ reaction. Angular distributions and polarization of γ-rays.

Entrance channel			Exit channel	Multipole transition	Angular distribution	Polarization ($\Theta_\gamma = 90°$)
S	L	J^π	J^π			
0	1	1^-	0^+	$E1$	$\sin^2\Theta_\gamma$	100%, parallel
0	2	2^+	0^+	$E2$	$\sin^2\Theta_\gamma\cos^2\Theta_\gamma$	100%, parallel
1	0	1^+	0^+	$M1$	Isotropic	Unpolarized
1	1	1^-	0^+	$E1$	$1+\cos^2\Theta_\gamma$	100%, perpendicular

iii) The most general Laue-Bragg law may be expressed by saying that the condition for a coherent scattering in a crystal is fulfilled when the recoil momentum $\boldsymbol{q}$ of the nucleus is equal to the reciprocal lattice vector $\boldsymbol{g}$. Let a relativistic electron have a momentum $\boldsymbol{p}$ along the direction which forms a small angle ϑ with the axis of the reciprocal lattice of a crystal having a lattice spacing a. Then

$$q = 2p\sin\vartheta \tag{5.4.2}$$

$$g = \frac{2\pi}{a}n , \tag{5.4.3}$$

where n is an integer number.

The recoil momenta in a high-energy bremsstrahlung process involving a single nucleus have their end points in a disk-shaped region, perpendicular to the primary electron momentum and displaced by δ from the origin of the momentum vector. The thickness of the disk is of the order of δ and its radius is of the order of 1 (in units of m_0). By using classical arguments only (Dia 68), one can deduce that the minimum momentum transfer to a single nucleus δ depends on the bremsstrahlung photon energy and is given by

$$\delta = \frac{1}{2E_1}\frac{K/E_1}{1-K/E_1} \quad (\text{in units of } m_0^{-1}) , \tag{5.4.4}$$

where K is the energy of the photon and E_1 the maximum energy of the bremsstrahlung spectrum (both in units of m_0). For $K/E_1 \ll 1$ one has $\delta \sim K/2E_1^2$ which thus increases with K/E_1.

In a perfect crystal, without thermal motion at all, only the bremsstrahlung processes are permitted for which $q = g$. This means that we have to superimpose the recoil-momentum region on the reciprocal lattice of the crystal and to take into account only the intersection zone of the disk-shaped momentum region with the reciprocal lattice plane perpendicular to the row of atoms which defines g. This intersection zone depends on δ and therefore on K/E_1.

Let us now fix the angle ϑ and look at the bremsstrahlung spectrum. Each time a K/E_1 value allows to have an intersection, a discontinuity appears as a peak in the bremsstrahlung spectrum. It is easy to show that the photon energies at which the discontinuities occur are given by the expression

$$\left.\frac{K}{E_1}\right|_n = \frac{1}{1+\left(2\frac{4\pi}{a}nE_1\theta\right)^{-1}} . \tag{5.4.5}$$

It is clear that in order to have this effect the recoil momentum must be of the order of the minimum reciprocal lattice vector. This condition is fulfilled in the bremsstrahlung of high-energy electrons. Moreover, the condition $\delta \ll |g_0| = 2\pi/a$ must be fulfilled in order to have a large effect. The most suitable crystal is diamond because of its small lattice spacing. Since $a = 3.56$ Å we have $2\pi/a = 6.8 \cdot 10^{-3}$, in units of m_0. For $E_1 = 1$ GeV one gets $\delta = 1.02 \cdot 10^{-3}$ for $K/E_1 = 0.8$ and $\delta = 1.09 \cdot 10^{-4}$ for $K/E_1 = 0.9$ in the same units. A diamond crystal is also very appropriate because of its high Debye temperature, which reduces the perturbation due to the thermal motion.

We have already pointed out in ii) that the high-energy bremsstrahlung photons are always linearly polarized. Moreover, if we consider the "recoil plane", determined by the direction of the primary electron and the recoil momentum of the nucleus $(\bar{p}_1, \bar{q})$, the state of linear polarization is parallel to this plane. For an amorphous target the whole bremsstrahlung is almost completely unpolarized. On the contrary in a crystal, the bremsstrahlung beam can have a net polarization with respect to a crystal plane, since only definite recoil momenta are allowed. This polarization has a maximum value when only one point of the reciprocal lattice contributes to the peak, while it is very low when several points are involved, because it becomes a combination of all the contributions weighted over the different effective $\bar{q} = \bar{g}$ values. Let us call as "incident plane" the plane $(\bar{p}, \bar{b}_1)$ between the primary electron momentum and the reciprocal lattice axis $\bar{b}_1$ and α the angle between the planes $(\bar{p}, \bar{b}_1)$ and $(\bar{b}_1, \bar{b}_2)$, where $\bar{b}_2$ refers to another lattice axis. By using the classical arguments and if the thermal vibrations are not taken into account, the polarization at the top of a peak of the bremsstrahlung spectrum due to only one point of the reciprocal lattice is given by the following expressions

$$P_{(\bar{q},\bar{p})} = \frac{2(1-K/E_1)}{1+(1-K/E_1)^2} \quad \text{in the "recoil plane"} , \tag{5.4.6}$$

$$P_{(\bar{p},\bar{b}_1)} = \frac{2(1-K/E_1)}{1+(1-K/E_1)^2} \cos 2\alpha \quad \text{in the "incident plane"} . \tag{5.4.7}$$

As a conclusion one can say that a rather high degree of polarization (0.5-0.7) can be obtained by selecting a large angle ϑ and a small angle α.

Generally a goniometer device allows the crystal to be rotated around a vertical and a horizontal axis in order to choose the most convenient conditions and to change the polarization vector of the beam parallel or perpendicular to the reaction plane. It can easily be demonstrated that the angular spread of the primary electron beam and the multiple scattering in the crystal do not affect the polarization. As a consequence, crystals as thick as usual radiators can be used.

iv) We have already discussed this process in section 5.2, since it is a method to obtain monochromatic photons. Moreover, in the Compton scattering of laser photons by high-energy electrons the secondary photon is also polarized, since the incident light is polarized. Following Hei 54 the differential cross section is given by

$$\frac{d\sigma}{d\Omega} = \frac{r_0^2}{4}\frac{k^2}{k_0^2}\left[\frac{k_0}{k}+\frac{k}{k_0}-2+4\cos^2\Theta\right] , \tag{5.4.8}$$

where $k_0(k)$ is the primary (secondary) photon momentum and Θ is the angle between the directions of polarization of the primary and secondary photons.

From (5.4.8) it is evident that the polarization of the emitted photon has a maximum value near the top of its energy spectrum, i.e., for the photon emitted in the direction of the electron. Let us consider the scattered radiation as composed of two orthogonal linearly polarized components. Denoting the directions of polarization of $\boldsymbol{k}_0$ and $\boldsymbol{k}$ by $\boldsymbol{e}_0$ and $\boldsymbol{e}$, respectively, we can choose the following two directions for $\boldsymbol{e}$

($\perp$): $\boldsymbol{e}$ perpendicular to $\boldsymbol{e}_0$

($\|$): $\boldsymbol{e}$ parallel to $\boldsymbol{e}_0$.

According to (5.4.8) the $\|$ component is always more intense than the $\perp$ component. If the incident light has polarization 1 the emitted-photon polarization can reach a value of 0.9 to 1, as has been obtained by FeG 80.

Now the question is how to determine the polarization of a photon beam and with what degree of precision. This experimental problem is rather important because the uncertainty of the photon polarization affects the measured cross section values. The first method consists in evaluating the polarization, taking into account all the experimental parameters that can affect its value. The second method consists in measuring it by means of a nuclear process or by means of an electromagnetic effect.

For high-energy photons the angular distribution of the electron pair production has been analyzed by BaB 62. The measurement of the polarization depends upon the fact that in electron pair production from linearly polarized photons, the electrons are emitted preferentially in the plane of polarization (i.e. the plane determined by the direction of the incident photon and the polarization vector). More details can be found in Dia 68.

In another case (TaM 60) the meson production process has been employed. Assuming the production of pions in the Δ-resonance region essentially due to a magnetic dipole transition, the production at right angles to the electric field vector of the photons is expected to be dominant. The precision of the polarization that one obtains by these experimental methods is not better than the precision given by the calculations. In both methods one must assume that the cross sections or/and the experimental conditions are well known. The measured values of the polarization were generally affected by an error $\Delta P/P$ of $10-20$ percent.

In CaC 78 the resonant scattering of γ-rays on single nuclear levels has been analyzed in order to measure the incoming-photon polarization. The 15.1 (1^+) level of ^{12}C has been used and the polarization P at this energy has been deduced by the measurement of the intensities $I^{\|}(\vartheta)$ and $I^{\perp}(\vartheta)$ of the scattered photons, since

$$I^{\|}(\vartheta) = K[(1-P)\cos^2\vartheta + P] , \tag{5.4.9}$$

$$I^{\perp}(\vartheta) = K[P\cos^2\vartheta + (1-P)] . \tag{5.4.10}$$

Here, $I^{\parallel}(\vartheta)$ and $I^{\perp}(\vartheta)$ are the intensities of the scattered photons at a scattering angle ϑ, when the angle between the scattering plane and the plane containing both the direction of the incident photon and its polarization vector is 0° and 90°, respectively. In the measurement the scattering angle $\vartheta = 90°$ has been chosen and the polarization P has been obtained with a very good precision of $\Delta P/P = 5 \cdot 10^{-3}$.

For what concerns the polarized neutron beams, a high polarization has been obtained by AmB 77 and AbB 85 by bombarding a deuterium target with a transversely polarized proton beam. Without describing the facility in detail, which one can find in AmB 77 and AbB 85, we give here a schematic description of the method. The transversely polarized proton beam was obtained using a superconducting precession solenoid in order to rotate the proton spin by 90° into the horizontal plane. Neutrons emitted by the deuteron target pass through two dipole magnets, which precess the transverse component of the neutron polarization into the vertical plane. Typically the neutron beam intensity was of the order of 10^6 s^{-1} with 50 - 60 percent polarization. More recently a polarized neutron beam in the neutron energy range $E_n = 19 - 50$ MeV and a polarization of 36 to 53 percent has been obtained using the $d(\vec{d},n)$ reaction (FiD 90). Also in the case of neutron beams the polarization is known with an accuracy of 10 – 20 percent.

6 Experimental results for two-particle break-up

In this section we shall review the present status of experiments on two-body photodisintegration of deuteron up to and above the Δ region. The results of different experiments will be discussed and a guideline will be given for a selection of the experiments to be used in the comparison of experimental data with theoretical results. Moreover, since a large variety of observables exists, which have been studied experimentally, we shall discuss which kind of measured quantities, having a smaller systematic error, can give more accurate information.

6.1 Total and differential cross sections

In most of the experiments on photodisintegration of the deuteron the differential cross section has been measured for several angles from which a more or less complete angular distribution has been deduced. The total cross section, except for a few cases, has been evaluated from the measured angular distribution.

As outlined in section 3.5 the most appropriate representation for the differential cross section in the cm system is the one in terms of Legendre polynomials (see (3.5.4) using here A_n instead of V_{00}^n)

$$\frac{d\sigma(E_\gamma,\Theta)}{d\Omega} = \sum_{n=0}^{\infty} A_n(E_\gamma)P_n(\cos\Theta) , \tag{6.1.1}$$

where Θ is the angle between the incoming-photon and outgoing-proton momenta in the cm system and E_γ is the lab photon energy.

Because of the orthogonality of the Legendre polynomials the coefficients A_n, as given by

$$A_n(E_\gamma) = \frac{2n+1}{2}\int_{-1}^{1} \frac{d\sigma(E_\gamma,\Theta)}{d\Omega}P_n(\cos\Theta)d(\cos\Theta) , \tag{6.1.2}$$

are independent from each other and their values do not depend on the truncation of the sum. Often the sum is truncated at $n = 3$ or $n = 4$. Other expansions like the one of DSw 59 and Par 64 (see appendix B) or just a simple power series in $\cos\Theta$ have also been used.

However, as pointed out in section 3.5 the coefficients in such an expansion are strongly correlated and their values depend on the choice of the maximal order included, i.e., on the truncation. Obviously for such an expansion it is always possible to express the coefficients as a linear combination of the A_n coefficients of the Legendre expansion. An example is given in appendix B.

From the expansion (6.1.1) one deduces

$$\sigma_{tot} = 4\pi A_0 . \tag{6.1.3}$$

It must be noted once more that (6.1.3) is always true and does not depend on the truncation of the sum. The precision of the total cross section, deduced from the angular distribution, depends not only on the precision of the measured differential cross sections but also and especially on the angular range in which the differential cross section has been measured.

Only in a few experiments the total cross section of the two-body photodisintegration of the deuteron has been directly determined either using a detector with a very large angular acceptance ($\sim 4\pi$) or measuring the total photon absorption cross section. Obvioulsy this last method is valid only below the pion photoproduction threshold where the only photon-induced reaction is the two-body emission.

In the total photon absorption measurements a heavy-water target has been used. In order to subtract the contribution of the nuclear photoabsorption by oxygen and of the atomic processes in the D_2O, normal- and heavy-water absorbers of equal effective length have been employed and the deuteron cross section was obtained by measuring the difference in photon absorption by the two targets. The magnitude of the deuteron disintegration cross section σ_D^{nucl} is much lower than the total cross section for photon absorption in water $\sigma_{H_2O}^{tot}$, e.g., at a photon energy of 20 MeV $\sigma_D^{nucl} \leq 10^{-3}\sigma_{H_2O}^{tot}$. Thus, for a cross section determination with a ± 5 percent accuracy, the relative difference $\Delta L/L$ in absorber lengths for H_2O and D_2O must be kept smaller than $5 \cdot 10^{-5}$.

In the experiment of AhE 74, in order to minimize any influence of a difference in the absolute absorber lengths, each absorber holder contained H_2O for about half of the measurements and D_2O for the other half. With the method just described the total deuteron photodisintegration cross section has been measured by BiK 85 using monochromatic γ-rays from the (n, γ) reactions at photon energies between 6 and 11.4 MeV and by AhE 74 employing a bremsstrahlung beam at photon energies of 15, 20 and 25 MeV. The same method was used by Col 51 at 2.62 and very recently by MoK 89 at 2.76 MeV γ-ray energy.

For a first general survey we show in fig. 6.1.1 the experimental total cross section for two-body disintegration of the deuteron between threshold and 400 MeV photon energy as measured in several laboratories and using different procedures. In fig. 6.1.2 we present in more detail the experimental total cross section above 60 MeV photon energy.

A large discrepancy exists between the experimental data mainly at energies above 100 MeV, where the difference in the absolute value can be as large as a factor 2. Now the question is: what is the origin of this large discrepancy? Is it due to the analysis, i.e., uncertainties in the fitting of the angular distribution or does it come from an inconsistency already present in the differential cross section measurements? In fig. 6.1.3 we show, as an example, the differential cross sections measured in several experiments for incoming photon energies $E_\gamma = 100$, 220, and 300 MeV.

It is evident from this figure that there is a large difference between the values of the differential cross section measured by some of the experiments and it is reflected in the absolute value of the total cross section evaluated from these data. On the other hand, in the whole energy range, starting from threshold up to high energies, even if we find large differences in the absolute value of the cross sections measured by various experiments, we do not find so much disagreement with respect to the relative behaviour of the total cross section with the energy, the position of the Δ-resonance and its width, and the relative behaviour of the angular distributions.

We can deduce, therefore, that the main reason for the disagreement of the absolute values is some systematic error which affects in the same way all the absolute values measured in one experiment. Typical sources of such systematic errors are the lack

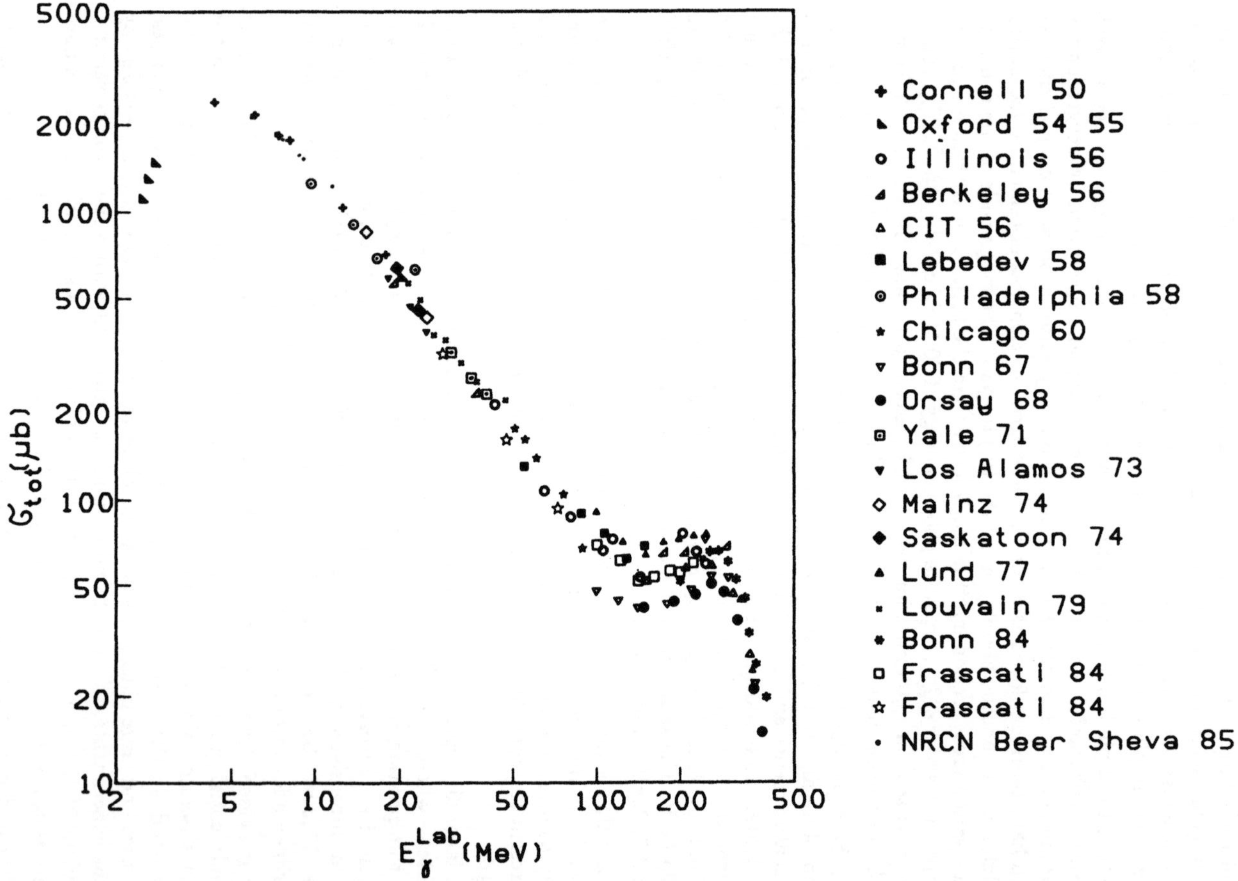

Fig. 6.1.1: Experimental data on total cross section for $d(\gamma,p)n$ as measured in several laboratories between threshold and $E_\gamma = 400$ MeV photon lab energy (from San 86).

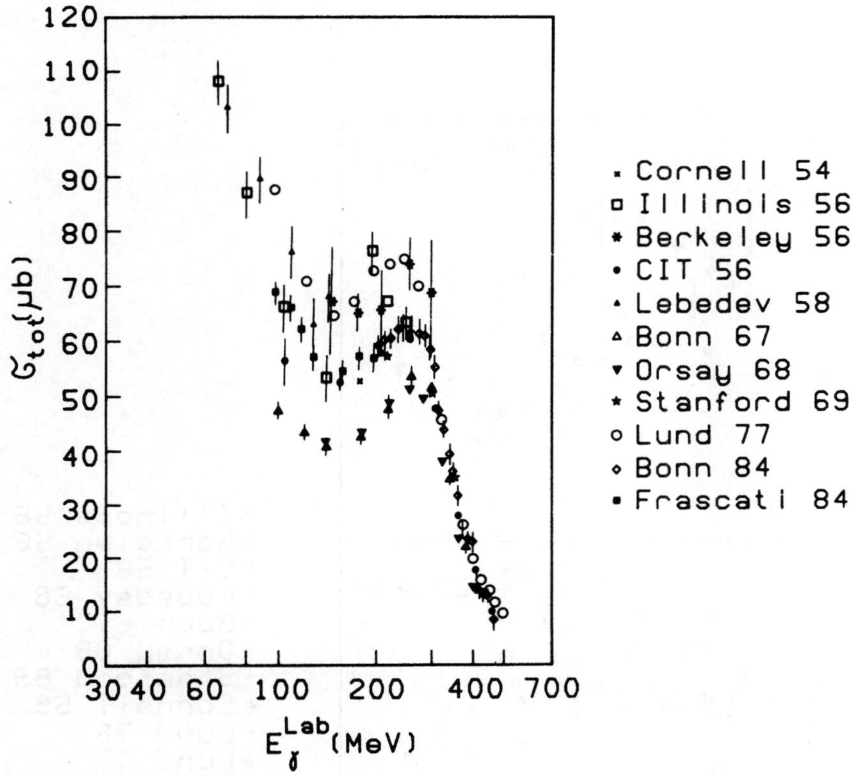

Fig. 6.1.2: Same as fig. 6.1.1 presented in more detail between E_γ = 60 and 500 MeV photon lab energy.

of a precise knowledge of the incoming photon flux, of the effective thickness of the deuteron target and of the absolute value of the detector efficiency. A deuteron target is a difficult target. Most of the experiments used either a gas target, deuterated water or paraffin and only a part of the experiments used a liquid deuteron target.

It is well known that for a gas or an enriched target it is not so easy to know the effective thickness. The effective thickness of a liquid target can be known with a much better precision mainly if one controls the bubble formation in the liquid due to a local power loss of the impinging beam.

For what concerns the detector efficiency it is difficult and sometimes impossible to make critical comments on the detector apparatus used in each experiment, since often the tests performed on the apparatus are not described in the publications. In any case the lack of an accurate knowledge of the effective thickness of the deuteron target and of the absolute detector efficiency cannot explain the large discrepancy one finds between the experimental results of various laboratories. On the contrary, this can surely be due to the uncertainty in the absolute number of photons employed in the experiment.

Most of the experiments used the continuous bremsstrahlung photon beam produced by the deceleration of fast electrons by a radiator. As we have already discussed in section 5.1, uncertainties in the knowledge of the precise shape of the radiation spectrum

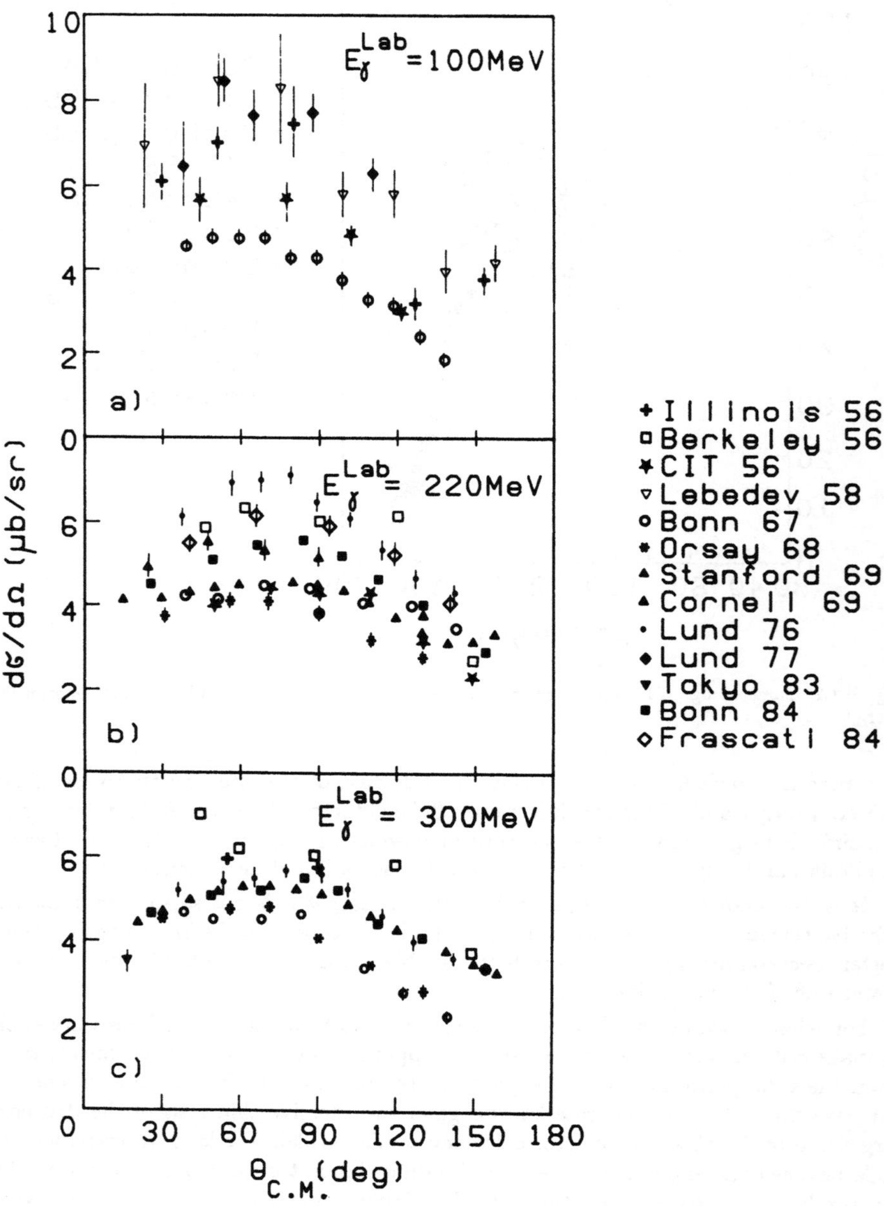

Fig. 6.1.3: Experimental data on differential cross section for $d(\gamma,p)n$ at photon lab energies E_γ = 100, 220, and 300 MeV as measuered in several laboratories.

can give a big error in the evaluation of the number of photons with a fixed energy. It must be noted that in most of the experiments the Schiff bremsstrahlung spectrum, without taking into account any multiple scattering distortion of the spectrum, has been used because of its simple analytic formulae.

Moreover, very often the bremsstrahlung beam has been collimated and the photons near the end-point of the spectrum have been employed for the experiment, where, according to table 5.1.1, the discrepancy between the theoretical values is much larger. As a conclusion we can say that the systematic error, coming from the lack of a precise evaluation of the incoming photon flux, can be the main source of the large discrepancy between the experimental results on the absolute cross section for the photodisintegration of the deuteron.

6.2 Information from analysis of all available experiments

Despite the systematic errors, which affect the absolute value of the measured cross sections, it is always possible to have information from the experimental results. The first way is to use relative quantities which are not sensitive to the absolute value of the cross section, i.e., the relative angular distribution, the ratio between two differential cross sections, etc. The second way is averaging the results of different experiments and doing a best fit of experimental data that takes into account both statistical and systematic errors, if known.

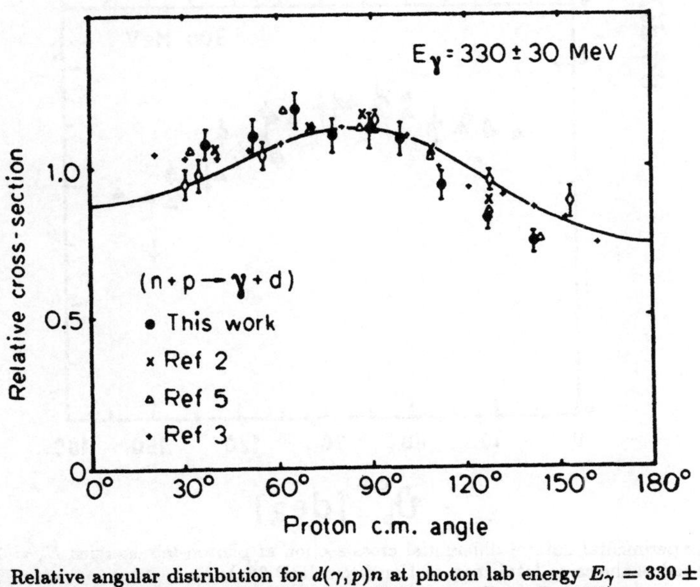

Fig. 6.2.1: Relative angular distribution for $d(\gamma,p)n$ at photon lab energy $E_\gamma = 330 \pm 30$ MeV from different experiments with total cross section normalized to 4π (from DoK 76).

As a first example we show in fig. 6.2.1 the relative angular distribution from different experiments at a photon energy $E_\gamma = 330 \pm 30$ MeV as taken from DoK 76. The data of each experiment have been normalized by requiring that the total cross section has a value of 4π. Furthermore, we show in fig. 6.2.2 – from RoD 89a – the differential cross section data at 100 MeV and 300 MeV as produced by the authors and after normalization to the same total cross section values. In both figures the results of different experiments appear to be in reasonable agreement with each other. Most of the measurements presented in figs. 6.2.1–2 come from the same experiments already presented in fig. 6.1.3 without any normalization of the data. It is clear that despite the large spread in absolute values measured in the different experiments, the relative angular distributions are rather similar.

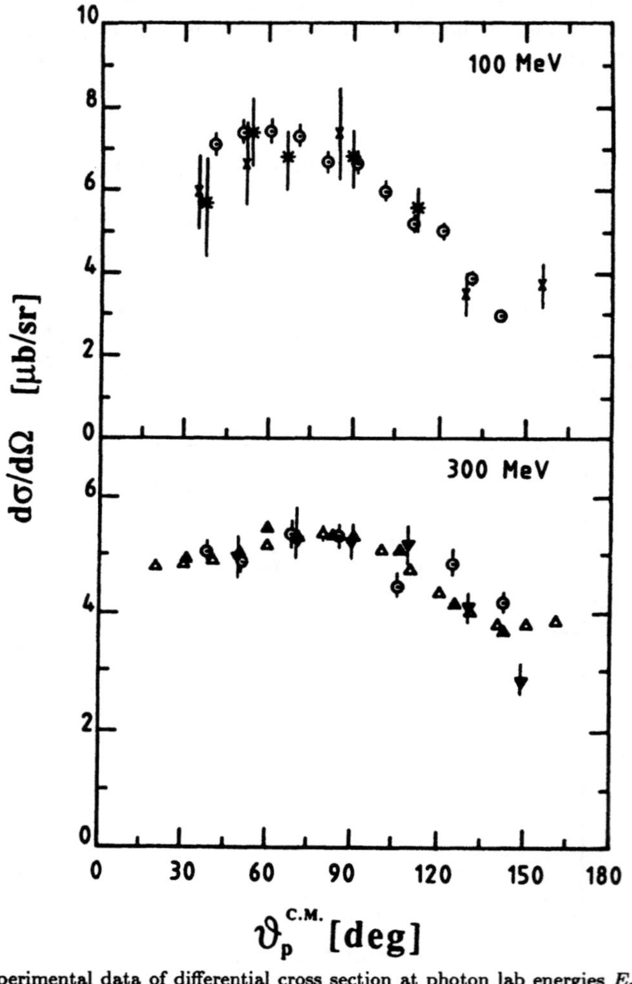

Fig. 6.2.2: Experimental data of differential cross section at photon lab energies E_γ = 100 and 300 MeV normalized to the same total cross section (from RoD 89a).

A second example can be the measurement of the ratio between photomagnetic and photoelectric cross section near the threshold. The theory of photodisintegration of deuteron predicts that for gamma-ray energies close to the threshold the photomagnetic effect dominates over the photoelectric one which, however, becomes increasingly important as the photon energy is increased. As a consequence the angular distribution of the emitted nucleon should be represented in the cm system by $a + b\sin^2\Theta$. The spherically symmetric contribution arises from the magnetic effect, while the $\sin^2\Theta$ term arises from the electric effect.

The ratio of the total photomagnetic to photoelectric cross section σ_m/σ_e is equal to $3a/2b$. Several experiments have been performed, mainly during the 1950's, in order to measure the ratio a/b. Most of them just counted the number of protons or neutrons emitted at 0° and 90° under the same experimental conditions. It follows that

$$\frac{a}{b} = \left(\frac{N_{90^\circ}}{N_{0^\circ}} - 1\right)^{-1} . \qquad (6.2.1)$$

This quantity is much less affected by systematic errors than the absolute values of a and b.

In table 6.2.1 the results of different experiments are reported together with the indication of the apparatus used in each experiment. We see that different experimental methods have been employed. Nevertheless the values of the ratio a/b are consistent among each other within the errors. It must be noted that some systematic errors can still affect the measurement of this value. They are due mainly to background contamination that can be appreciable in the experiments which detect neutrons and to the scattering of photoneutrons inside the heavy-water target. They give an apparently larger a/b ratio, since they contribute to the isotropic term. This can be verified looking at table 6.2.1, where the values obtained by experiments, in which the neutrons were detected, are generally higher than the other ones.

Tab. 6.2.1 Experimental data on the ratio a/b.

Photon energy (MeV)	a/b	experimental procedure	references
2.51	0.41 ± 0.03	b	BiH 51
2.51	0.40 ± 0.015	a	BiB 51
2.62	0.26 ± 0.08	b	GrH 45
2.62	0.24 ± 0.006	a	BiB 51
2.76	0.18 ± 0.04	a	Las 49
2.76	0.196 ± 0.024	b	Gen 49
2.76	0.211 ± 0.008	b	Mei 49
2.76	0.205 ± 0.05	b	HaW 49a,b
2.76	0.165 ± 0.005	a	BiB 51
2.76	0.193 ± 0.014	c	SmB 87

a: proton detection with proportional counter filled with D_2-gas
b: neutron detection from D_2O
c: proton and neutron detection in coincidence from deuterated anthracene scintillation crystal

A fit of the existing cross section data has been obtained by De Pascale et al. (DPG 82) for photon energies up to 120 MeV. Later, another fit was given at higher energies

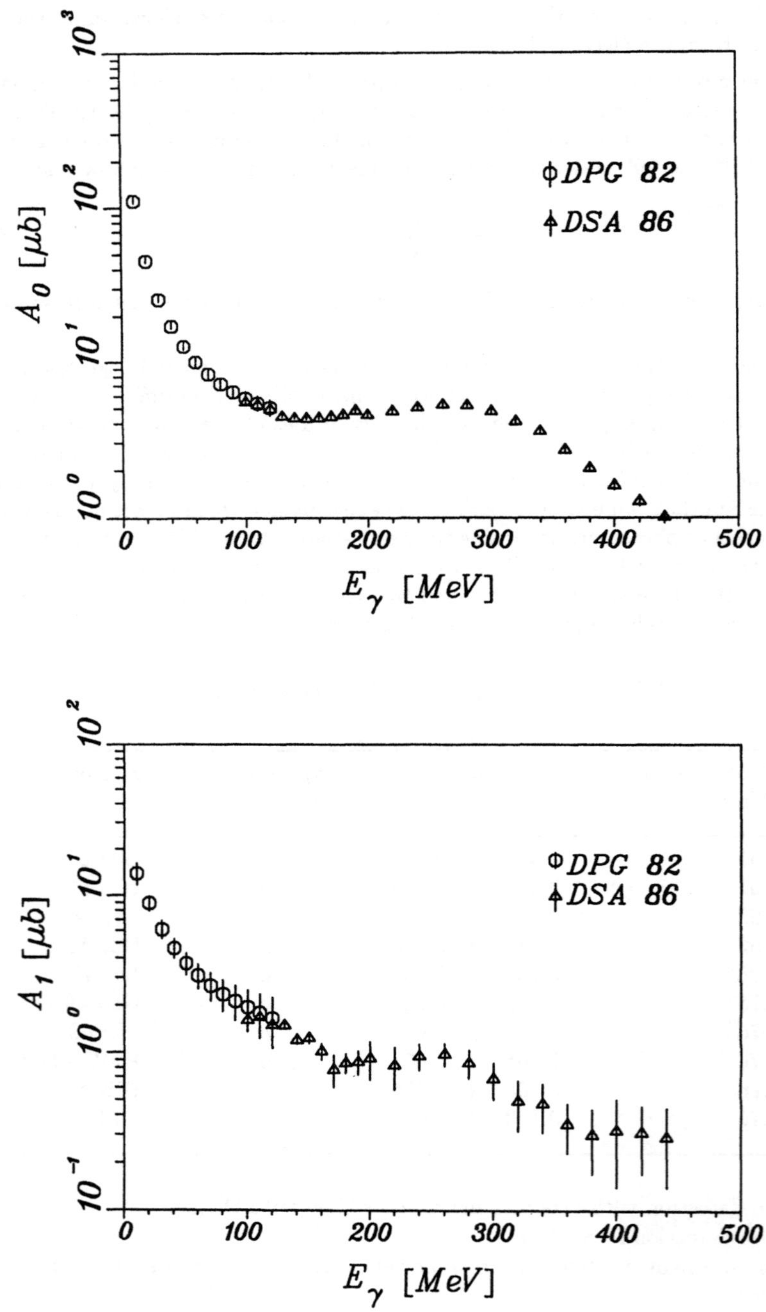

Fig. 6.2.3: Best fit values of Legendre coefficients A_0 and A_1 of differential cross section data.

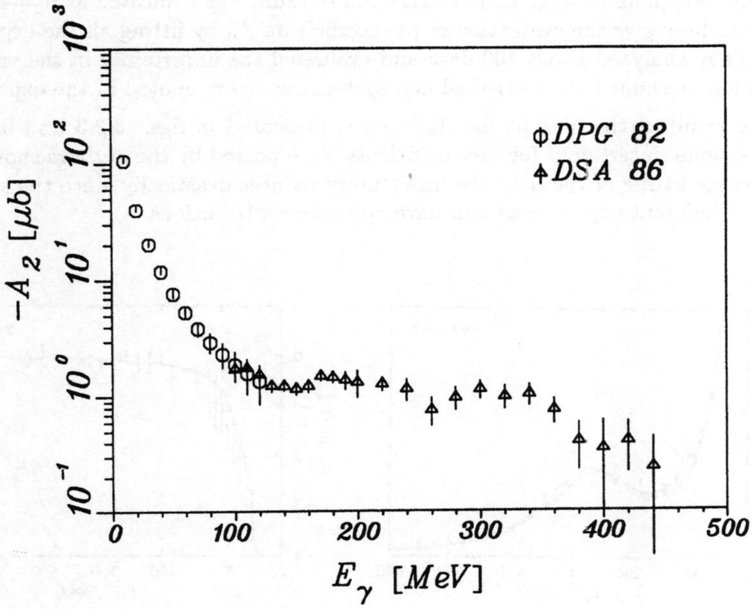

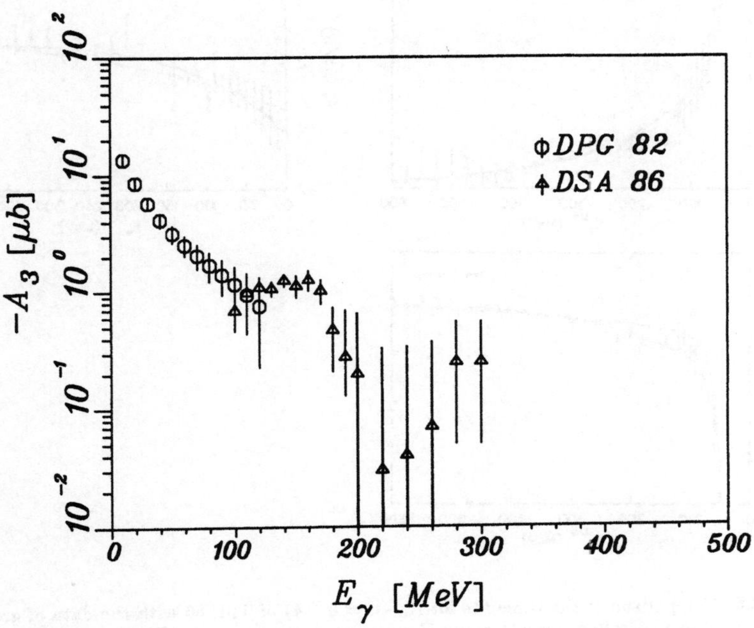

Fig. 6.2.4: Best fit values of Legendre coefficients A_2 and A_3 of differential cross section data.

by De Sanctis et al. (DSA 86). The differential cross section was written in terms of Legendre polynomials as given in (6.1.1) and the sum was truncated at $n = 4$. In DPG 82 the authors give the evaluation of the coefficients A_n by fitting all the experimental data. They analyzed about 400 data and evaluated the uncertainty in the value of A_n taking into account both statistical and systematic errors quoted in the experiments.

The result of the best fit for $A_0 - A_3$ is presented in figs. 6.2.3 and 6.2.4. The relative small uncertainty for the coefficients A_n reported by the authors shows that in the average fitting of the data the uncertainty reduces drastically since the systematic errors of different experiments can have opposite contributions.

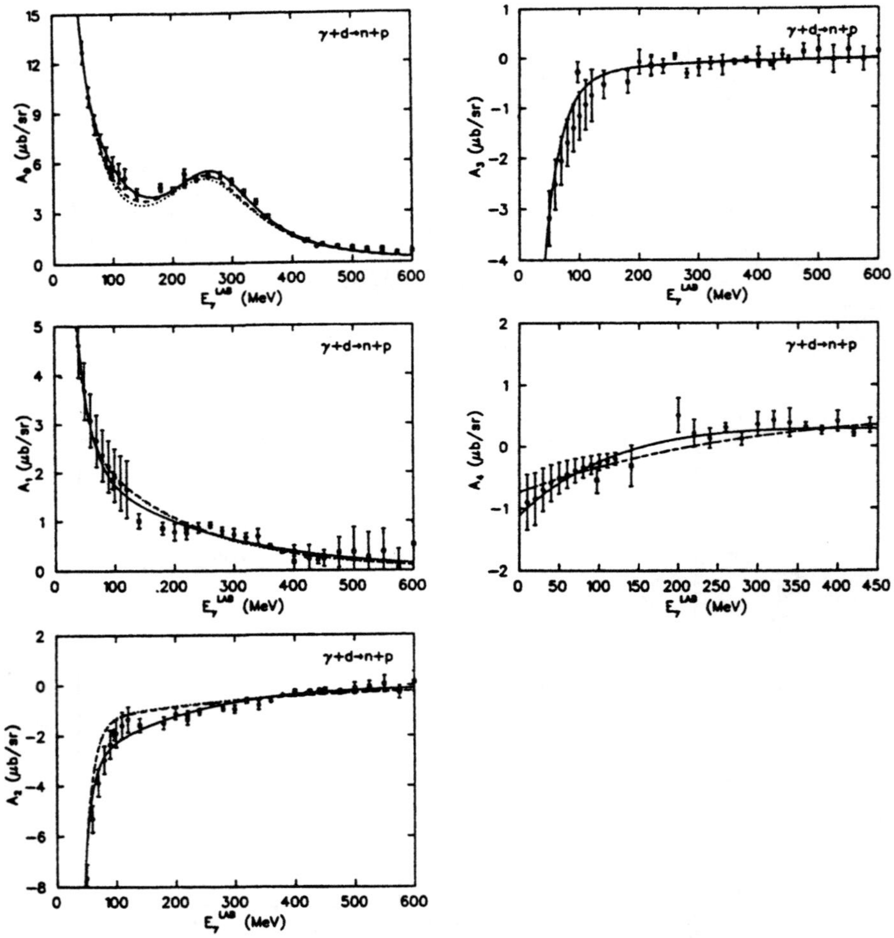

g. 6.2.5: Comparison of the three fits for A_n ($n = 0 - 4$) of ThF 86 with the data of group I. Full rve fit to group I, dashed curve to group II and dotted curve to group III. Experimental data: open xes DoR 77, open circles AlA 83 and ArG 84, closed boxes MeH 85, DSC 85, Hut 84 and CaD 84 d closed circles fit of DPG 82 at lower energies (from ThF 86).

In ThF 86 and RoD 89b an attempt was made to fit an analytical expression for the Legendre coefficient A_n as a function of the energy:

$$\bar{A}_n(E_\gamma) = c_{n1}e^{c_{n2}E_\gamma} + c_{n3}e^{c_{n4}E_\gamma} + \delta_{n0}\,\frac{c_5 + c_6 E_\gamma}{1 + c_8(E_\gamma - c_7)^2}\,. \tag{6.2.2}$$

The coefficients are listed in ThF 86 and RoD 89b. These phenomenological fits $\bar{A}_n(E_\gamma)$ allow a simple evaluation of the experimental differential cross section as function of lab photon energy E_γ and cm angle Θ.

Thorlacius and Fearing (ThF 86) made a fit to three different groups of data. Group I consists of all the more modern data. Group II consists of all available medium-energy data, while in Group III the data of two experiments DoR 77 and SoC 69, which are significantly higher than most other data, have been excluded. In this analysis the fit of DPG 82 was used for the lower energies and data included in that fit were dropped so as to avoid double counting.

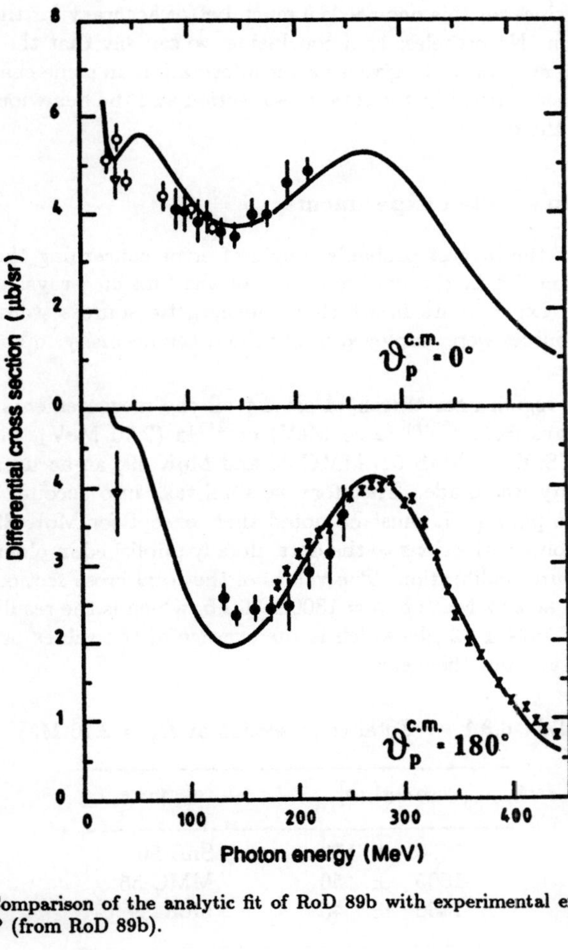

Fig. 6.2.6: Comparison of the analytic fit of RoD 89b with experimental excitation function at $\Theta_p = 0°$ and 180° (from RoD 89b).

In fig. 6.2.5 the results of the three fits are shown together with the data of Group I and the fit of DPG 82. Also in this analysis the uncertainty of the differential cross section decreases sensitively by averaging all the data. RoD 89b derived the absolute value of the total cross section from selected experiments carried out with monoenergetic photons and then used the whole available angular distribution measurements, suitably normalized, to obtain a phenomenological representation of the differential cross section up to 400 MeV.

In fig. 6.2.6 from RoD 89b the excitation functions at 0° and 180° obtained by the fitting procedure are compared with existing data. Also in this analysis it is evident that a better representation of the angular distribution can be obtained by the fitting, while the averaging procedure cannot reproduce so well some details like the differential cross section at extreme angles.

However, we would like to remark that these analytical fits can be useful only when an approximate value of the cross section is needed, since the averaging and the smoothing over a large energy range can give only a less accurate result for the A_n. This means that in any case they cannot be used for a detailed comparison of theory with experiment, since for this one needs a much better accuracy for the absolute value of the cross section. Nevertheless as a conclusion we can say that the fit methods are useful and consistent in order to give a better information on some characteristic gross features of the cross section like the total cross section and the behaviour of the angular distribution with the energy.

6.3 Results from selected experiments

Assuming that the biggest probable source of error concerning the absolute value of the cross section lies in the determination of the flux of γ-rays, we will select in the following the experiments in which monoenergetic sources were used or where bremsstrahlung sources were employed but without the necessity to know theoretically the detailed shape.

In the energy region near threshold almost all the measurements have been performed using γ-rays from ^{208}Tl (2.62 MeV) or ^{24}Na (2.76 MeV), but only for very few experiments (SnB 50, MaB 54, MMC 55 and MoK 89) an accurate calibration of the source intensity was made. Therefore we shall take into account only the results presented in these papers. It must be noted that, except for MoK 89, all the values are the result of some corrections to the data already published in older papers, mainly concerning the source calibration. The values of the total cross section near threshold are the following: at 2.62 MeV: $\sigma_{tot} = 1300 \pm 29$ μb, which is the result of MaB 54. At 2.76 MeV: $\sigma_{tot} = 1474 \pm 32$ μb, which is the average of the values presented in table 6.3.1, taking into account their errors.

Tab. 6.3.1 Total cross section at $E_\gamma = 2.76$ MeV

σ_{tot} [μb]	reference
1430 ± 110	SnB 50
1505 ± 50	MMC 55
1456 ± 45	MoK 89

In the photon energy region between 4 and 18 MeV we have selected the measurements of the total cross section performed using γ-rays arising from capture reactions induced by protons (Hou 50, WäY 51 and BaC 52) or by neutrons (BiK 85). In the last experiment the total cross section was measured using the absorption method described in section 6.1. In table 6.3.2 we report the values of the total cross section together with the γ-ray energies.

Tab. 6.3.2 Experimental results on the total cross section

E_γ[MeV]	σ_{tot} [μb]	references
4.45	2430 ± 170	BaC 52
5.97	2162 ± 99	BiK 85
6.14	2190 ± 100	BaC 52
7.25	1882 ± 11	BiK 85
7.39	1840 ± 150	BaC 52
7.60	1803 ± 16	BiK 85
7.64	1810 ± 28	BiK 85
8.14	1800 ± 130	BaC 52
8.80	1586 ± 11	BiK 85
9.00	1570 ± 36	BiK 85
11.39	1257 ± 36	BiK 85
12.50	1040 ± 100	BaC 52
14.80	870 ± 200	WäY 51
17.60	720 ± 150	Hou 50
17.60	710 ± 150	WäY 51
17.60	770 ± 90	BaC 52

In the photon energy range between 15 and 30 MeV we have selected two experiments, in which the total cross section was measured. The first one, AhE 74, used the absorption method and the deuteron cross section was obtained from the difference in photon absorption by heavy and normal water, as discussed in section 6.1. Moreover a bremsstrahlung photon beam was employed, measuring the intensity of photons of a given energy directly by a Compton spectrometer. In this experiment a systematic error of the absolute cross section value can arise only from a difference in the atomic cross section for normal and heavy water.

In the other selected experiment SkS 74 the equivalent photon cross section has been extracted from deuteron electrodisintegration measurements, assuming a virtual photon spectrum in the long-wave-length approximation containing only the $E1$ multipole. So the main systematic error, concerning this experiment, comes from the assumed virtual photon spectrum.

A more recent experiment BeI 86 measured the total cross section between 15 and 75 MeV photon energy, using an approximately 4π-proton detector with an efficiency of almost 1 and monochromatic photons by Compton back-scattering of laser light off electrons circulating in a storage ring. The beam energy spectrum was analyzed by a magnetic pair spectrometer and the beam intensity was monitored by an NaI(Tl) scintillator counter of well-known efficiency. In this way the absolute number of photons of a given energy was known with a good accuracy.

In the photon energy region below 40 MeV we find also measurements of the inverse n-p capture cross section from Tud 67, BoB 79, DuL 85 and StK 85. As already discussed in section 5.3, from this cross section, using the detailed balance theorem, the cross section for photodisintegration of the deuteron can be deduced. The whole angular range of the deuteron (0° to 180° cm) is constrained in the laboratory system to a narrow forward cone with an opening angle of a few degrees. In the experiments Tud 67, BoB 79, DuL 85 and StK 85 the total cross section was measured, as the deuterons emitted in the whole cone were detected. Moreover, the capture cross section was deduced by comparing the number of deuterons with the number of protons scattered at a defined angle and normalizing the n-p capture cross section to the n-p scattering differential cross section. As a consequence the accuracy of the capture cross section depends on the uncertainty of the n-p elastic cross section, as discussed in section 5.3.

The results of all the experiments at $E_\gamma \leq 40$ MeV, mentioned in this section, are reported in fig. 6.3.1. The good agreement between the data is quite surprising, considering the fact that they were obtained with different methods. In fig. 6.3.1 the best fit of DPG 82 for $\sigma_{tot} = 4\pi A_0$ as discussed in section 6.2 is also presented. The data of the selected experiments are nicely reproduced by the values $4\pi A_0$ of DPG 82. This shows, as already discussed in section 6.2, that the average fitting of all the data reduces drastically the uncertainty due to the systematic errors. As a consequence, one can take the results of DPG 82 like experimental values having good accuracy.

At higher energies we can select only few experiments. They have been performed recently by determining the cross section of deuteron photodisintegration through the measurement of the inverse n-p capture cross section (MeH 84, 85, CaD 86, HuC 87, Edw 89), employing monochromatic photons from tagged bremsstrahlung (BaE 83, ArG 84, AnW 88, Wal 89), from positron annihilation in flight (DSA 86, LeA 89), or from Compton back-scattering (DPG 85).

At higher energies another problem arises from the use of a photon beam with a continuous spectrum. Above the threshold for pion photoproduction, protons associated with the incoherent pion production are also emitted from the deuteron. The competing reactions are

$$\begin{aligned} \gamma + d &\rightarrow \pi^\circ + p + n \\ &\rightarrow \pi^- + p + p \,. \end{aligned}$$

For a fixed proton energy and angle, the energy of the photon responsible for the two-body deuteron photodisintegration is also fixed. But there exists a limiting photon energy above which the pion photoproduction can contribute to the counting rate of these protons. Since the amount of this contribution depends on the proton direction and energy a distorted angular distribution and an absolute cross section with a large systematic error will be measured, if the difference between the energy of the photons employed in the experiment and the bremsstrahlung end-point energy is too large. As a consequence at energies above the pion production threshold, it is essential to use monochromatic photons in order to identify the two-body reaction unambiguously by imposing kinematic constraints.

We shall present also the data of one experiment which, even if performed using the continuous spectrum of a bremsstrahlung beam, presents an accuracy on the knowledge of the photon flux of few percent (Owe 90). The bremsstrahlung was not collimated, a medium-Z radiator (aluminum) was used and the proton spectrum up to 120 MeV below the bremsstrahlung endpoint energy was analyzed. These experimental conditions, as

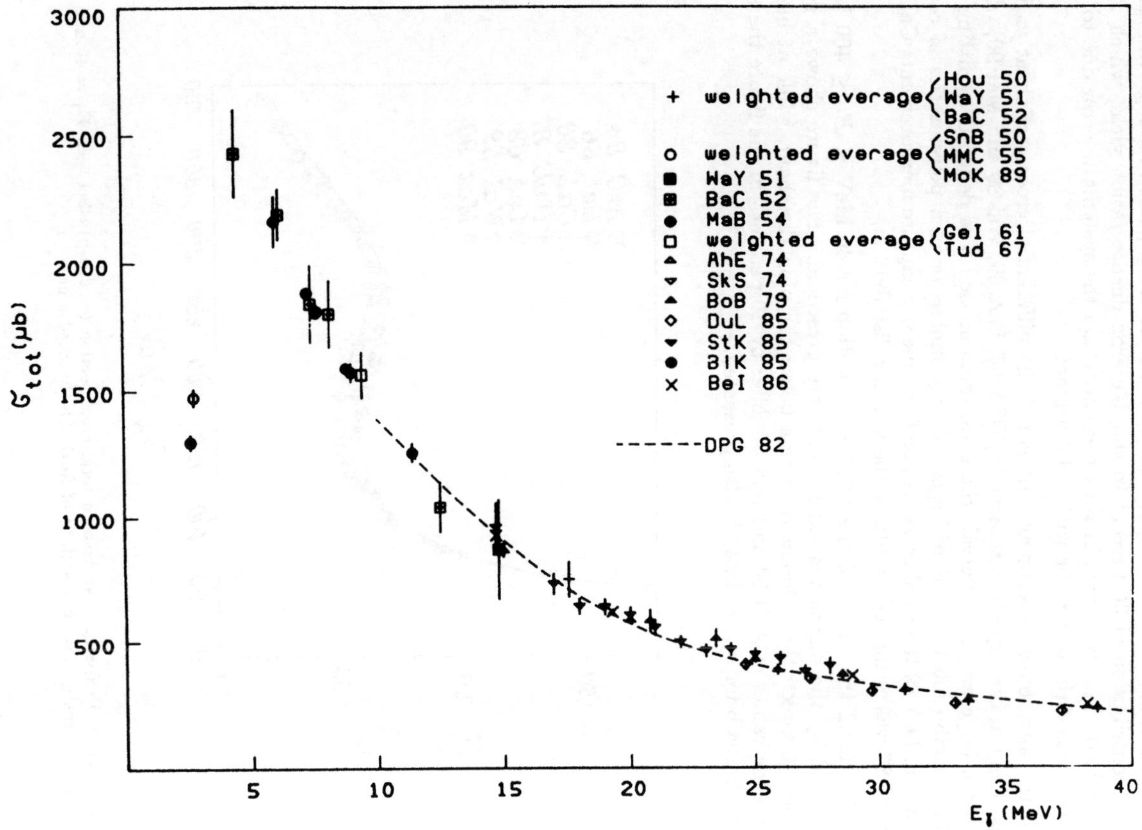

Fig. 6.3.1: Selected experimental data on total cross section for $d(\gamma,p)n$ between threshold and $E_\gamma = 40$ MeV photon lab energy together with fit of DPG 82.

discussed in section 5.1, reduce drastically the systematic error concerning the photon-intensity knowledge. Moreover, tests were performed in the experiment to check the validity of the assumed bremsstrahlung spectrum.

The main problem of this experiment is that the deuteron target was placed close to the bremsstrahlung radiator, so that the residual electrons passed through the target giving a contribution of 30 percent to the deuteron disintegration yield, which must be subtracted. This was done in the experiment and the systematic error due to this correction was estimated to be within ±3 percent.

All the mentioned experiments measured the differential cross section at various angles. Only in the experiments ArG 84, DSA 86, LeA 89, Wal 89 and Mat 90, which explored a large angular interval, the total cross section was deduced by fitting the angular distribution in terms of Legendre polynomials or in a power series of $\cos\Theta$. Moreover in LeA 89 the total cross section from the n-p capture measurements CaD 86 and HuC 87 was evaluated by fitting the angular distribution.

In fig. 6.3.2 the values of the total cross section for 40 MeV $\leq E_\gamma \leq$ 400 MeV as obtained by the mentioned experiments are presented. The figure shows a good agreement between the different data sets below the pion threshold, while at higher energies the values of ArG 84 and LeA 89 differ by 10 percent. In the figure the best fit of RoD 89a,b for $\sigma_{tot} = 4\pi A_0$ is also presented.

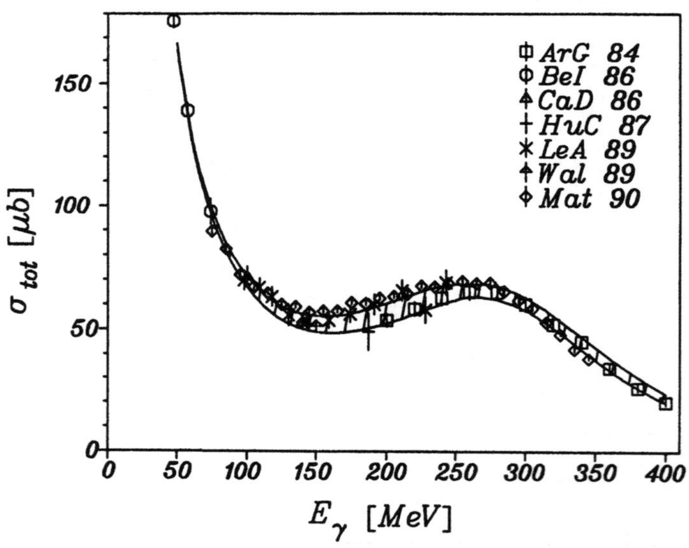

Fig. 6.3.2: Selected experimental data on total cross section for $d(\gamma,p)n$ between E_γ = 40 and 400 MeV photon lab energy together with fit of RoD 89b (hatched area).

It must be noted that below 140 MeV the best fit of DPG 82 for A_0 was assumed in RoD 89b, while at higher energies the total cross section was derived from experiments carried out with monoenergetic photons only. As a consequence, in the Δ region the fit of RoD 89b is just the weighted average of the ArG 84 and LeA 89 data, as it appears clearly in fig. 6.3.2.

For what concerns the angular distribution we present in figs. 6.3.3–4 the differential cross sections for $E_\gamma = 20$ MeV, 100 MeV, 140 MeV, 220 MeV, and 400 MeV as measured by the experiments selected in this section together with the fits of DPG 82 and DSA 86. In some cases the data of *n-p* capture experiments have been transformed, using the formulae of section 5.3 in order to obtain the photodisintegration differential cross section.

In fig. 6.3.3 the data at $E_\gamma = 20$ MeV of FiD 89 are multiplied by a factor 1.16 for the following reason. In this paper the absolute differential cross sections were obtained by normalizing the values to the total cross section of the *n-p* capture. This value was deduced from the total photodisintegration cross section, applying the principle of detailed balance. Unfortunately, for the total photodisintegration cross section the data of WhH 58 and BaC 73 were used. These experiments were performed employing bremsstrahlung photon beams. We have reevaluated the total photodisintegration cross section by averaging the data of selected experiments (AhE 74 and SkS 74). The result is 16 percent higher than the value assumed by FiD 89 in their table 1.

Moreover, in fig. 6.3.3 we have scaled the data for $E_\gamma = 100$ MeV of MeH 85 and CaD 86 by an appropriate factor, in order to take into account that the original data were taken at slightly different photon energies. The scaling factors were evaluated taking into account the energy dependence of the total cross section.

Looking at figs. 6.3.1–4 we see that a big improvement has been obtained by selecting those experiments, where the intensity of the employed photons are known with a good accuracy. Below the pion production threshold a good agreement exists between the data on the total and differential cross sections. At higher energies the discrepancy between the experimental results is drastically reduced. Nevertheless we see that at photon energies above the pion threshold the data set derived from these experiments lies within a band of relative width (10 – 20 percent) which increases with the energy.

Before concluding this section we present in figs. 6.3.5 the energy dependence of the experimental differential cross sections at $\Theta^{cm} = 0°, 180°$ and $90°$, respectively. In this figure we have plotted the values measured directly at those angles in the direct or the inverse process. In the direct reaction $d(\gamma,p)n$, in order to detect the protons at 0° or 180° with respect to the incoming photon beam, a magnet system was used for extracting the protons from the beam and to transport them from the target to the detector. In HuZ 6, ZiG 86, DGr 88 and DGZ 89 the proton detectors were placed in the focal plane of a magnetic spectrometer, positioned behind the deuteron target. In LeD 87 and LeA 89 the deuteron target was inserted in the centre of a circular dipole magnet and the ray tracing of protons from the target to the detectors was reconstructed by a Monte Carlo simulation. For what concerns monitoring of the photon beam, the intensity of photons was measured by a Compton spectrometer in HuZ 76 and ZiG 86 and by a pair spectrometer in LeD 87 and LeA 89.

In DGZ 89 a bremsstrahlung beam was employed and the energy beam flux was monitored with an NBS-P2 ionization chamber. Even if generally this monitoring method can introduce a large systematic error with respect to the absolute value of the cross section, we report the results of DGZ 89, since the experimental conditions and some tests performed in this experiment (DGr 88) allow a systematic accuracy of a few percent as was the case for the experiment Mat 90. The bremsstrahlung beam was not drastically collimated and as a consequence the Schiff spectrum integrated over angles was assumed. Moreover the photons in the plateau region of the bremsstrahlung

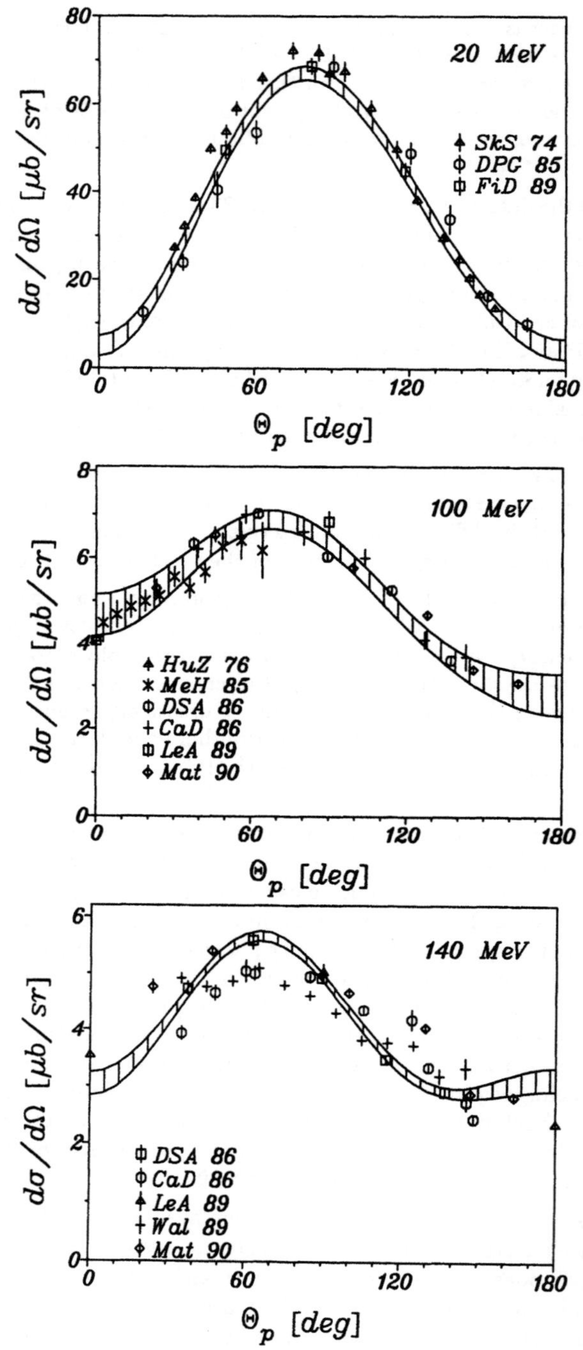

Fig. 6.3.3: Selected experimental data on differential cross section at photon lab energies $E_\gamma = 20$ MeV together with fit of DPG 82 (hatched area) and 100 and 140 MeV together with fit of DSA 86 (hatched area).

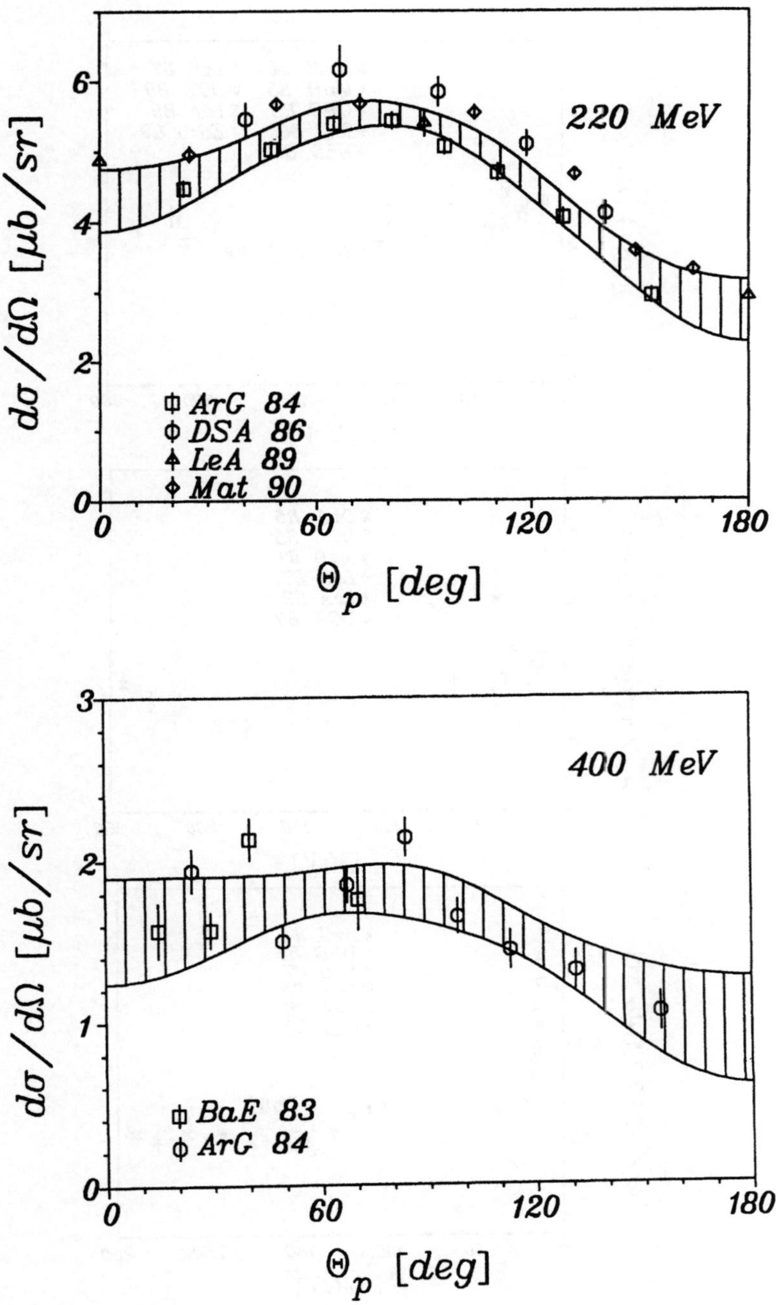

Fig. 6.3.4: Selected experimental data on differential cross section at photon lab energies E_γ = 220 and 400 MeV together with fit of DSA 86 (hatched area).

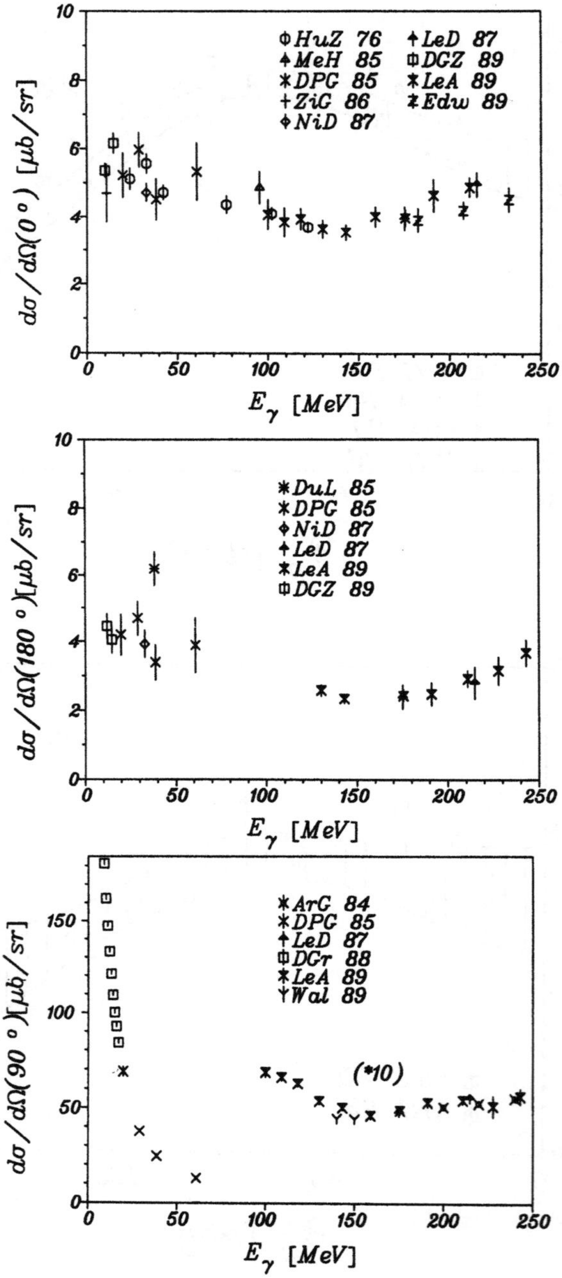

Fig. 6.3.5: Selected experimental data on differential cross section at 0°, 180°, and 90° cm angles. For 90° the values were multiplied by 10 above $E_\gamma = 70$ MeV.

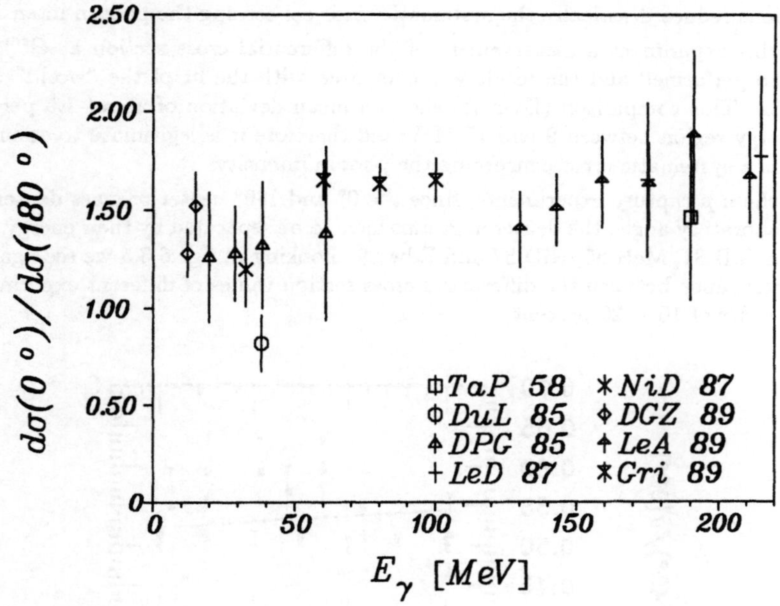

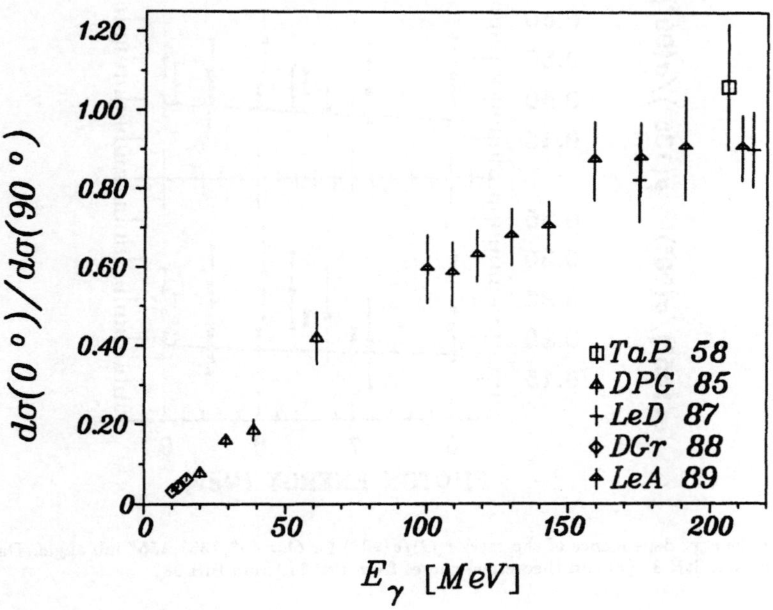

Fig. 6.3.6: Experimental data on cross section ratios $d\sigma(0°)/d\sigma(180°)$ and $d\sigma(0°)/d\sigma(90°)$, all cm angles.

intensity spectrum were employed. As we have discussed in section 5.1, these two conditions reduce drastically the systematic error concerning the photon intensity.

In this experiment a measurement of the differential cross section at $\Theta^{cm} = 90°$ was also performed and the result was compared with the fit of the "world" data of DPG 82. This comparison (DGr 88) shows a mean deviation of about 1.5 percent in the energy region between 9 and 17 MeV and therefore it is legitimate to assume this value as a systematic error concerning the photon intensity.

In the *n-p* capture experiments, since the 0° and 180° center-of-mass deuterons go to 0° laboratory angle, the deuteron groups have to be separated by their energy, as was done in GiB 81, MeH 85, NiD 87 and Edw 89. Looking at fig. 6.3.5 we see again that the discrepancy between the differential cross section values of different experiments is of the order of 10 – 20 percent.

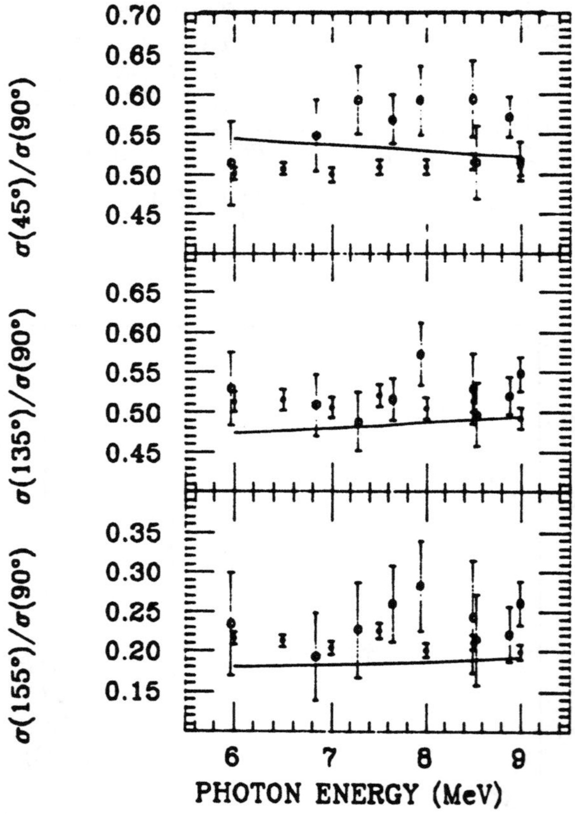

Fig. 6.3.7: Energy dependence of the ratio $\sigma(\Theta)/\sigma(90°)$ for $\Theta = 45°, 135°, 155°$ lab angle. Data from BiB 88 (o) and StH 87 (•) and theoretical curves from Par 64 (from BiB 88).

The effect of systematic errors, which contribute in a different way to each experiment, is much less in the ratio of two differential cross section values, as already discussed in section 6.2. For this reason we present in fig. 6.3.6 the ratios $d\sigma(0°)/d\sigma(180°)$

and $d\sigma(0°)/d\sigma(90°)$ where the two differential cross sections were measured in the same experiment. We see that the agreement among the data of different experiments is now much better, even if it must be noted that the ratio is affected by a statistical error larger than the one of each differential cross section value. This means that in order to obtain a ratio of two differential cross sections with high accuracy it is necessary first to perform measurements at each angle with very good statistics and then to minimize all the systematic errors which are angle-dependent.

This was done by StH 87 in a recent experiment where the ratios of the $d(\gamma, n)p$ cross sections at 45°, 135° and 155° to that at 90° (all lab angles) were measured below $E_\gamma = 18$ MeV. The statistical errors of the differential cross section ratios were of the order of 5 percent. Moreover, the precise measurement of the energy dependence of the neutron detector efficiency, the careful determination of the reaction angle and of the angular acceptance of the detector and the accurate evaluation of the neutron multiple scattering effect allowed to reach a total uncertainty (including both the statistical and systematic errors) of a few percent.

In fig. 6.3.7 the results of StH 87 for E_γ between 6 and 9 MeV are compared with the ones of a more recent experiment of BiB 88 in the same photon energy range. Also in BiB 88 the relative angular distribution was measured, but with much larger statistical uncertainty than in StH 87. Fig. 6.3.7 shows a rather good agreement between the results of the two experiments for the $\sigma(135°)/\sigma(90°)$ and $\sigma(155°)/\sigma(90°)$ ratios while for the $\sigma(45°)/\sigma(90°)$ ratio the data of BiB 88 lie systematically above the data of StH 87. In this figure only the statistical errors are reported. Since the disagreement appears only at the forward angle, we can deduce that a systematic error depending on the angle or on the neutron energy must be responsible. One possible source could be a different background contribution or an error in the evaluation of the efficiency for the 45° detector.

6.4 Experimental results for polarization observables

It is a well-known fact that polarization observables allow a further and much more detailed analysis of the process under study compared to the differential cross section alone. Because polarization observables contain a much richer information on the dynamics of the system than attainable without beam and/or target polarization and without polarization analysis of the particles in the final state. The reason for this is the fact that in contrast to the differential cross section, which is a sum of the absolute squares of the t-matrix elements, these polarization observables contain interference terms of the various reaction amplitudes in different combinations and, therefore, may be more sensitive to small amplitudes and to small contributions of interesting dynamical effects like, e.g., relativistic contributions and subnuclear degrees of freedom.

As shown in section 3.4 the spin structure of the deuteron photodisintegration reaction requires 12 complex helicity amplitudes in order to characterize completely the $\gamma d \to pn$ process. Hence 23 different observables have to be measured as a function of the photon energy and the proton cm angle. On the other hand one has to keep in mind that the study of polarization phenomena is experimentally quite involved and poses a great challenge to the experimentalist. Only in recent years considerable progress has been achieved in the development of polarized beams, oriented targets and particle polarimeters (Mey 85a,b).

As a consequence only few experiments have been performed to investigate single polarization quantities like beam or photon asymmetry Σ^l (using linearly polarized photons), target asymmetry T (using a vector and tensor polarized deuteron target) and recoil nucleon polarization P_y. More recently some data from double-polarization experiments performed with linearly polarized photons combined with a recoil nucleon polarization measurement, and from n-p capture of polarized neutrons have become available (VeK 81, BaB 86). In any case, compared with the number of observables the number of experiments is small and does not allow any systematic and complete analysis.

In the following we shall present the results concerning the photon asymmetry Σ^l, limiting ourselves to the asymmetry at 90° in the cm system, as a function of the photon energy, investigated by several authors, and to the angular distribution of the photon asymmetry at 60 MeV, which has been measured at different laboratories. Moreover, we shall show some results concerning the neutron polarization P_y. We have already discussed in section 5.4 the methods to obtain polarized beams and the precision obtainable for the polarization measurement. Therefore, we shall limit ourselves to presenting the data and to discussing their precision.

We shall now express the polarization observables Σ^l and P_y in terms of measured quantities. The cm differential cross section for the deuteron photodisintegration by linearly polarized γ-rays can be written in the form (see (3.4.10))

$$\frac{d\sigma(\Theta)}{d\Omega} = I_0(\Theta) + P_l^\gamma I_1(\Theta)\cos 2\Phi = I_0(\Theta)[1 + P_l^\gamma \Sigma^l(\Theta)\cos 2\Phi] , \qquad (6.4.1)$$

where Θ is the angle between the emitted nucleon and photon momenta in the cm system and Φ is the angle between the polarization and reaction planes (see fig. 3.2.1). P_l^γ represents the degree of linear polarization of the photon beam and

$$\Sigma^l(\Theta) = \frac{I_1(\Theta)}{I_0(\Theta)} \qquad (6.4.2)$$

determines the photon asymmetry of the differential cross section.

We define now $Y(\Theta, \Phi)$, the experimental yield at the angles Θ and Φ, that is the ratio of photonucleon events to the γ-ray flux, and $R(\Theta)$ the ratio between the yields measured for $\Phi = 90°$ and $\Phi = 0°$ under the same experimental conditions

$$R(\Theta) = \frac{Y(\Theta, 90°)}{Y(\Theta, 0°)} . \qquad (6.4.3)$$

Then one can easily deduce that the quantity $\Sigma^l(\Theta)$ is given by

$$\Sigma^l(\Theta) = \frac{1}{P_l^\gamma}\frac{1 - R(\Theta)}{1 + R(\Theta)} . \qquad (6.4.4)$$

The differential cross section for polarized n-p capture may be expressed (Par 64) as (see 3.4.24)

$$\frac{d\sigma(\Theta)}{d\Omega} = I_0^c(\Theta)[1 + P_t A_y(\Theta)\sin\Phi] , \qquad (6.4.5)$$

where P_t is the transverse polarization of the nucleon and again Θ is the angle between the emitted photon and nucleon momenta in the cm system and Φ the angle between the polarization and reaction planes (see fig. 3.2.2). $A_y(\Theta)$ determines the asymmetry of the differential cross section and is called analyzing power.

To evaluate experimentally $A_y(\Theta)$ two photon detectors are placed at a polar angle Θ on the left and right sides of the incoming beam. If one defines $N_L^{\uparrow(\downarrow)}$ as the number of counts in the left detector for beam spin up (down) and $N_R^{\uparrow(\downarrow)}$ the number of counts in the right detector for beam spin up (down), then the analyzing power can be evaluated experimentally (HuC 87) using the formula

$$A_y(\Theta) = \frac{\epsilon}{\overline{P}_n}, \tag{6.4.6}$$

where

$$\overline{P}_n = \frac{1}{2}(P_n^{\uparrow} - P_n^{\downarrow}) \tag{6.4.7}$$

$$\epsilon = \frac{\sqrt{\kappa}-1}{\sqrt{\kappa}+1} \tag{6.4.8}$$

$$\kappa = \frac{N_L^{\uparrow} N_R^{\downarrow}}{N_L^{\downarrow} N_R^{\uparrow}}. \tag{6.4.9}$$

Let us consider the inverse photodisintegration process and call $P_y(\Theta)$ the nucleon polarization for unpolarized incident photons in the direction perpendicular to the outcoming nucleon and photon momenta, where Θ is the cm angle of the nucleon. As already discussed in section 3.4 the outcoming nucleon polarization in $d(\gamma, N)N$ and the corresponding polarization observable in the n-p capture process are related to each other by

$$A_y(\Theta) = P_y(\Theta). \tag{6.4.10}$$

From the expressions (6.4.4) and (6.4.9) it appears that both $\Sigma^l(\Theta)$ and $A_y(\Theta)$ (or $P_y(\Theta)$), containing only ratios of yields, can be determined without knowing the absolute values of the γ-ray flux and of the target thickness. The only requirement is that these quantities must be the same for the measurements concerning the two polarization states. This means that $\Sigma^l(\Theta)$ and $A_y(\Theta)$ are not affected by systematic errors related to those quantities but only by the systematic errors due to lack of precise knowledge of the polarization degree of the beam, to different contributions of the background in each detector at different angles Φ and to different responses of each detector.

In fig. 6.4.1 we present the experimental values of $\Sigma^l(\Theta)$ at cm angle $\Theta = 90°$ for photon energies between 10 and 600 MeV, measured by several groups. The polarized photon beam employed in each experiment was obtained with one of the methods already discussed in section 5.4. They are: the technique of the off-axis bremsstrahlung selection in Liu 65 and KrS 88, the use of γ-rays from the ^{3}H$(p,\gamma)\,^4$He reaction emitted at $\Theta_\gamma = 90°$ in DBJ 80, the Compton scattering of laser light off high-energy electrons in DBF 81 and DPG 85, the method of coherent bremsstrahlung of electrons in a diamond single crystal in BaB 67, GoZ 79, GoZ 82, BaG 83, BaB 86, AdA 88, and VnG 88.

In fig. 6.4.2 we show the photon asymmetry $\Sigma^l(\Theta)$ measured at $E_\gamma = 60$ MeV by different experiments using linearly polarized photons obtained again with the different methods already mentioned. Figs. 6.4.1 and 6.4.2 show a rather good agreement between the data, since the maximum discrepancy is of the order of 10 percent.

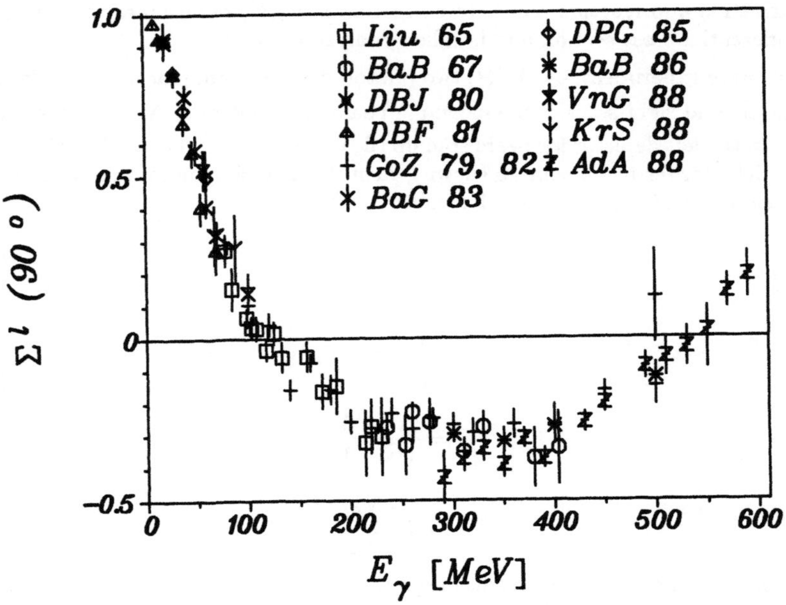

Fig. 6.4.1: Experimental data on photon asymmetry Σ^l at cm angle $\Theta_p = 90°$ between $E_\gamma = 10$ and 600 MeV.

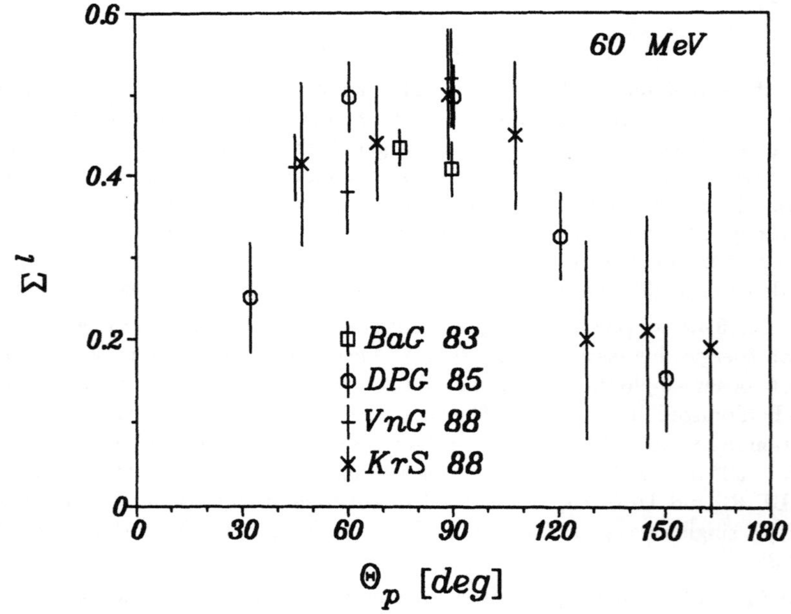

Fig. 6.4.2: Experimental data on angular distribution of photon asymmetry Σ^l at $E_\gamma = 60$ MeV.

For what concerns the analyzing power and neutron polarization measurements, we show in fig. 6.4.3 the results obtained with polarized n-p capture by FiD 89 along with the results of the polarization measurement of the outgoing neutrons of photodisintegration experiments with unpolarized photons from BeD 63, NaF 72, Dro 76, HoS 83 and SoK 87.

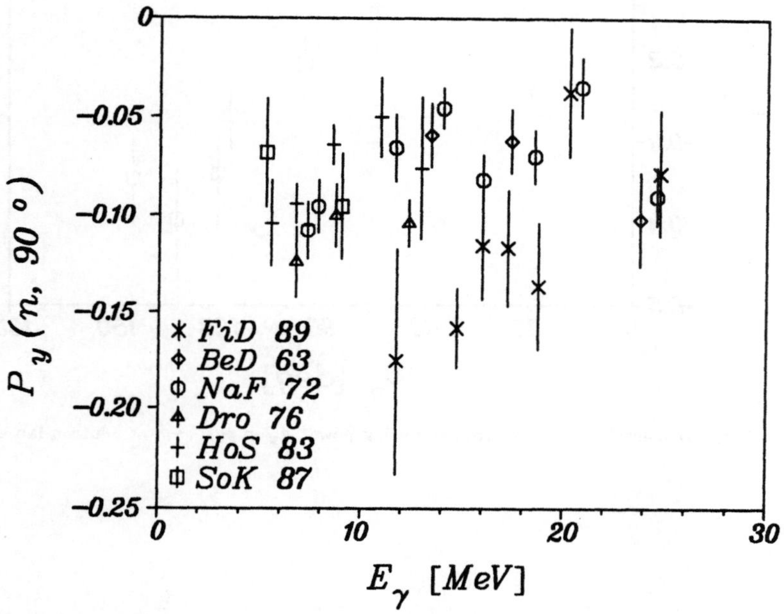

Fig. 6.4.3: Experimental data on neutron polarization P_y at cm angle $\Theta_n = 90°$ between $E_\gamma = 5$ and 25 MeV photon lab energies.

The data of $P_y(\Theta)$ presented in fig. 6.4.3 refer to $\Theta^{\rm cm} = 90°$ and to the photon energy range between 5 and 25 MeV. In this case the discrepancy between the data of different experiments is worse than that for the $\Sigma^l(\Theta)$. But it must be noted that the statistical errors of FiD 89 are large (see table 4 of FiD 89). Moreover, the sources of systematic uncertainty in the measurements of $P_y(\Theta)$ or $A_y(\Theta)$ are much larger than those in the measurements of $\Sigma^l(\Theta)$.

Finally in fig. 6.4.4 we show the values of the analyzing power $A_y(\Theta)$ for $E_n = 270$ MeV as measured by CaD 84. The errors are only those due to counting statistics. The uncertainty due to the determination of the neutron beam polarization was estimated to be independent of the energy and of a magnitude of about ± 0.025.

We conclude this section saying that few measurements of polarization observables exist at present. The uncertainty of the ensemble of data is generally within 20 percent like in the case of the differential and total cross section, even if the error sources are different. In the case of polarization observables they are mainly statistical errors due to the low intensity of polarized beams presently available and to the fact that these quantities have been obtained from the difference of counts of the same order of magnitude.

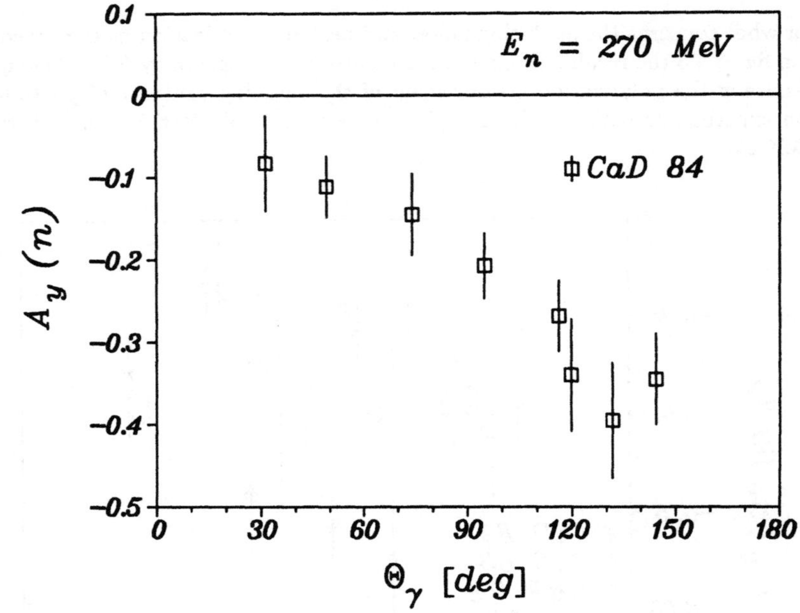

Fig. 6.4.4: Experimental data on neutron analyzing power A_y in $p(\vec{n},\gamma)d$ at neutron lab energy E_n = 270 MeV.

7 Comparison of theory and experiment

In this section we are finally going to confront experimental data on total cross section, angular distributions and polarization observables with theoretical results in order to establish the present status of our knowledge of this important process. We hope this will enable us to set directions and guidelines for future experimental and theoretical studies, which kind of experiments are needed and what accuracy should be aimed for, where theoretical studies should be extended, on what they should focus and whether some basic assumptions need to be modified. For the comparison we shall use mostly the data of experiments selected in sections 6.3 and 6.4, the relative quantities and the best fits discussed in section 6.2.

In most cases we shall use theoretical results based on the non-relativistic approach since it is the most elaborate one, but relativistic corrections will be considered as well. Furthermore, we shall also discuss theoretical results of the other approaches outlined in section 4. In order to distinguish the various ingredients, which enter into the theoretical description, we introduce the following nomenclature:

(i) "one-body" for pure one-body current contributions, i.e., no use of Siegert operators,

(ii) "normal (N)" for one-body current plus meson-exchange currents as incorporated in the Siegert operators (see 4.1.19), which we shall often call MEC of Siegert type,

(iii) "N+MEC" for normal plus explicit meson-exchange currents beyond the Siegert operators,

(iv) "N+MEC+IC" for additional isobar configurations and MEC contributions involving isobar configurations,

(v) "N+MEC+IC+SO or RC" for additional spin-orbit current or further relativistic corrections to the one- and two-body currents.

Since a large variety of calculations using different realistic potentials like Paris (LaL 80), Reid soft core (Rei 68) etc. exist, which give very similar results, we shall often present results for the Bonn r-space potential of Machleidt, Holinde, and Elster (MaH 87, Mac 89) based on a recent calculation by Schmitt and Arenhövel (ScA 89b) in which the different OBE approximations (r-space, q-space, time-ordered) of the Bonn full-potential model have been studied in detail. We have chosen the Bonn r-space potential as a typical example for a realistic N-N potential and refer to ScA 89b for details. But where necessary, we shall also discuss the potential model dependence in detail and present calculations of other groups and other approaches as well.

7.1 Total cross section and sum rules

Before confronting experimental data with theoretical results we briefly discuss the various theoretical ingredients contributing to the total cross section. For a general survey we show in fig. 7.1.1 the theoretical total cross section for two-body disintegration of the deuteron between threshold and 500 MeV photon energy as obtained with the Bonn r-space potential. Beyond the maximum at $E_\gamma = 2\epsilon_B \approx 4.5$ MeV the cross section drops rapidly with increasing energy up to π-threshold by a factor of about 50. Then a broad shoulder around $E_\gamma = 260$ MeV indicates the excitation of the $\Delta(1232)$ resonance followed by a further fall-off.

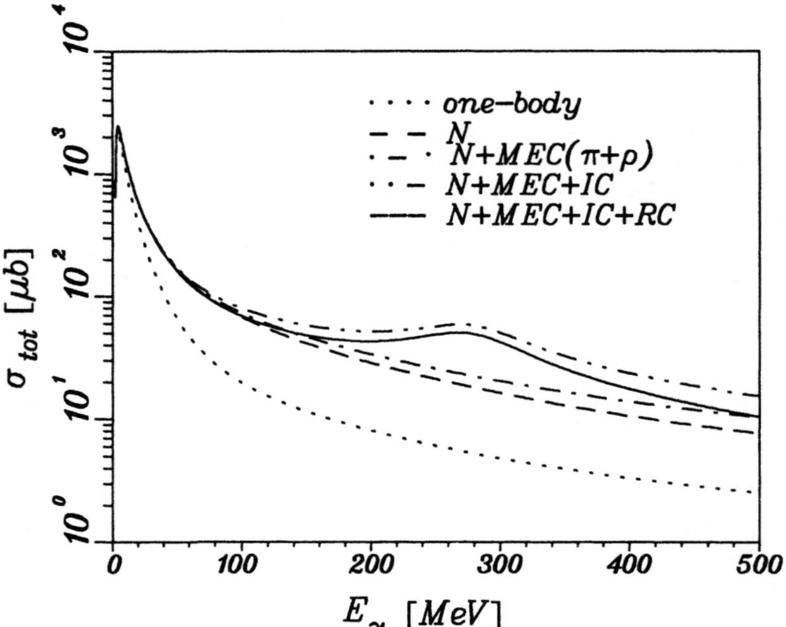

Fig. 7.1.1: Total photodisintegration cross section calculated for the Bonn r-space potential with various contributions from meson-exchange currents (MEC), isobar configurations (IC) and relativistic corrections (RC).

The different contributions from MEC, IC and RC are shown separately in fig. 7.1.1. The strong underestimation of the cross section, if only the pure one-body current is considered, demonstrates very clearly the overwhelming importance of MEC above about $E_\gamma = 20$ MeV. But as already mentioned in section 4.1, most of it is included in the Siegert operators of "N" which constitutes the classical theory. Taking this classical theory as a reference, we show in fig. 7.1.2 the relative strength of the various contributions. Above $E_\gamma = 80$ MeV the one-body part gives a constant 30 percent contribution. Additional explicit MEC beyond the Siegert operators from π- and ρ-exchange show an almost linear relative increase with energy, e.g., at $E_\gamma = 200$ MeV of about 20 percent.

Isobar configurations lead at low energy, except right at threshold, only to a small renormalization due to the presence of $NN(1440)$ and $\Delta\Delta$ components in the deuteron wave function. But above $E_\gamma = 100$ MeV they become of increasing importance, in particular in the final state, and dominate the total cross section for $E_\gamma = 240 - 320$ MeV. In this energy region they contribute by about $60 - 65$ percent to the total cross section, mostly through the $M1$ N-Δ excitation. On the other hand, the remaining $35 - 40$ percent of "N" being essentially MEC of Siegert type clearly show that the normal contribution cannot be neglected.

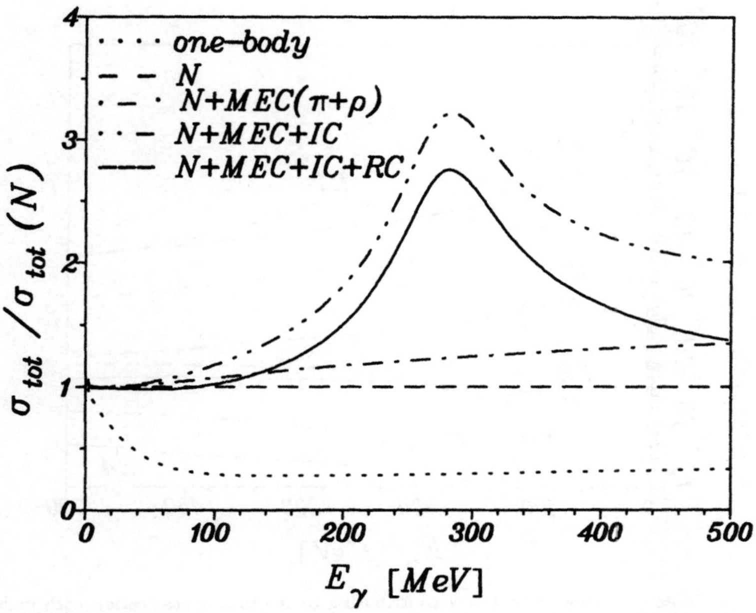

Fig. 7.1.2: Relative contributions of MEC, IC and RC to the total cross section with respect to the normal theory (N) for the Bonn r-space potential.

Above $E_\gamma = 40$ MeV the relativistic corrections, mainly from the spin-orbit current become of increasing importance and lead to a sizeable reduction of the total cross section, e.g., at $E_\gamma = 200$ MeV by about 15 percent.

In fig. 7.1.3 we show the contributions of electric and magnetic multipoles to the total cross section up to $L = 4$. The dominance of $E1$ up to π-threshold is readily seen. Only in the Δ-excitation region $M1$ becomes more important due to the strong $M1$ N-Δ transition. Here also $M2$ appears sizeable being an order of magnitude larger than $E2$ (LeA 87). The very rapid decrease of the higher multipole contributions justifies very well the restriction to $L \leq 4$ in most calculations. For $L = 1$ and 2 we show in fig. 7.1.4 the different contributions from MEC, IC and RC separately. It is obvious that contrary to what is often said the strongest evidence for MEC below π-threshold is seen not in $M1$ but in $E1$ transitions which, however, is usually hidden by the use of the Siegert operators. But also in $M1$ one has relatively large contributions both from MEC and IC. For the higher multipoles $E2$ and $M2$ the influence of MEC is small. This is a result of the suppression of the medium and short range parts of the current operators by the higher multipole operators. With respect to IC, they show up significantly also in $M2$, whereas they influence little the electric multipoles. Relativistic corrections, which are largest in $E1$, lead in general to a reduction, the size of which increases with energy.

After this general survey of the theoretical results, we shall now compare experiment and theory for the various energy regions, i.e., the low-energy region up to $E_\gamma = 10$ MeV, the intermediate-energy region up to the π-threshold, the Δ-excitation region between $E_\gamma = 200$ MeV and 400 MeV and the high-energy region.

The low-energy region is shown in fig. 7.1.5. It is well described by the classical

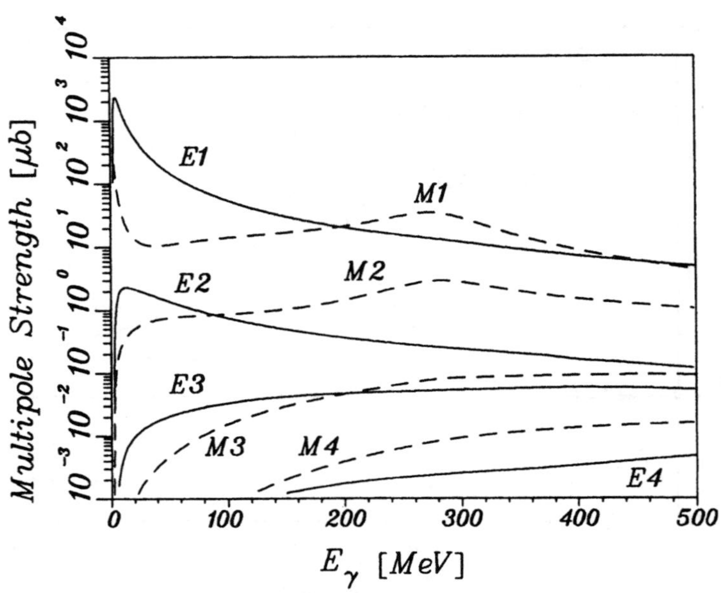

Fig. 7.1.3: Multipole strengths up to $L = 4$ contributing to the total cross section with inclusion of MEC, IC and RC for the Bonn r-space potential.

theory (N), and all other contributions from explicit MEC, IC, and SO are of minor importance except close to threshold. Comparison with the one-body contribution shows that already in the maximum of the total cross section ($E_\gamma = 4.5$ MeV) MEC contribute by about 8 percent. The steep rise right at threshold is caused by the $M1$ transition to the antibound $^1S_0$-resonance as is illustrated in fig. 7.1.6 where we show the $M1$ and $E1$ cross sections separately. The agreement with experimental data is quite satisfactory (see also RuP 89).

For a long time it has been noted that at threshold a discrepancy exists between experiment as measured in the inverse reaction of thermal n-p capture (CoW 65) and the classical theory using the effective-range approach (BeL 50, Aus 53, AuR 60). From (2.21) and (3.4.25) one obtains for the capture cross section right at threshold in the shape-independent approximation of the effective-range theory

$$\sigma_c^{th} = \pi \frac{e^2}{M^2}(\mu_p - \mu_n)^2 \frac{\epsilon^2}{\alpha k} \frac{(1 - \alpha a_s + \frac{1}{4} a_s (r_s + r_t)\alpha^2)^2}{(1 - \alpha r_t)} . \tag{7.1.1}$$

Inserting the experimental values it predicts a cross section $\sigma_c^{th} = 305$ mb, which is about 10 percent too small compared to the experimental value $\sigma_c^{exp} = 334.2 \pm 0.5$ mb (CoW 65). A thorough discussion of the theoretical uncertainties from shape-dependent corrections is given by Noyes (Noy 65). Estimating these uncertainties he arrives at a theoretical value $\sigma_c^{th} = 302.5 \pm 4.0$ mb. Furthermore, he pointed out the inadequacy of dispersion-theoretical approaches (SaG 62, BoC 63, Sko 64), which predicted a larger normal cross section.

Various attempts to explain this discrepancy by explicit π-MEC contributions (Fra 58, Sko 64, KaL 66, AdC 70, KaL 72) failed because the important pair or contact current had been left out. Riska and Brown (RiB 72), including this pair or contact

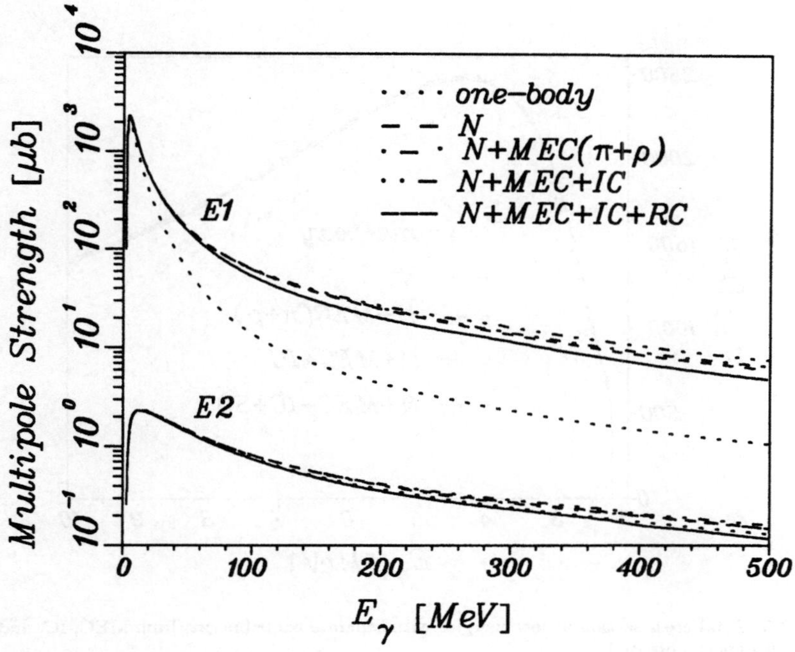

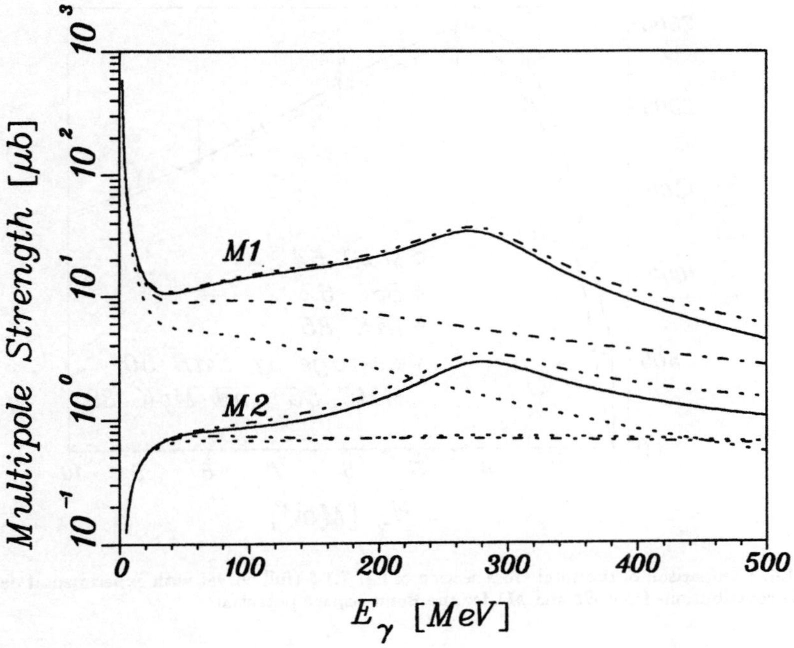

Fig. 7.1.4: Contributions of MEC, IC, and RC to multipole strength up to $L = 2$ for the Bonn r-space potential.

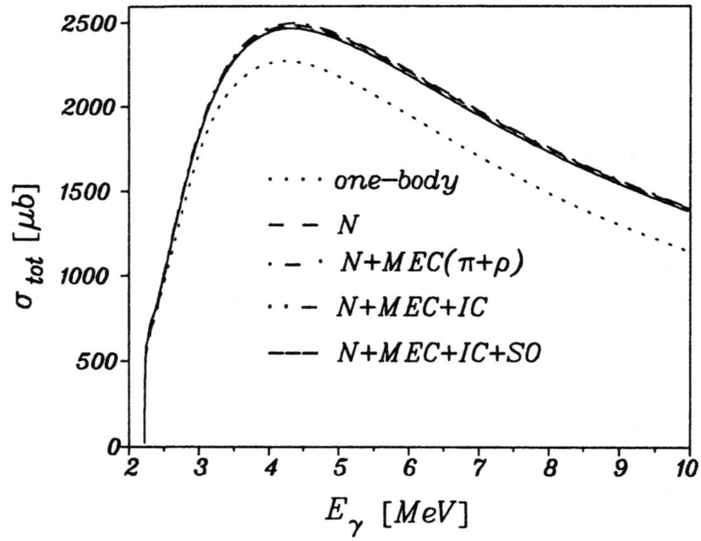

Fig. 7.1.5: Total cross section at low energies with separate contributions from MEC, IC, and SO for the Bonn r-space potential.

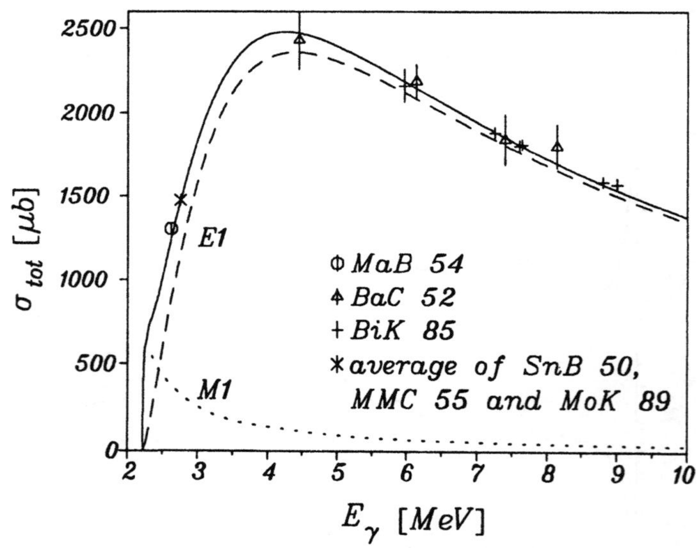

Fig. 7.1.6: Comparison of the total cross section of fig. 7.1.5 (full curve) with experimental data and separate contributions from $E1$ and $M1$ for the Bonn r-space potential.

current and pointing out the importance of S-D transitions, were the first to show that the complete π-MEC enhances the normal cross section by about 6.5 percent. Adding the estimate of the Δ contribution from Str 64 of about 3 percent they could explain

Tab. 7.1.1 Relative contributions to $M1$ transition matrix elements to ${}^1S_0$ state from interaction effects and capture cross sections for radiative thermal n-p capture without (σ_c^N) and with (σ_c^{tot}) interaction effects

Potential	relative contributions to $M1$-transition (in percent) π-MEC	ρ-MEC	IC	π-MEC-IC	ρ-MEC-IC	IC_{ren}	total	$a_s[fm]$	$\sigma_c^N[mb]$	$\sigma_c^{tot}[mb]$	ref.
RSC (Rei 68)	2.99		1.12	0.39		0.9952	3.95	-17.10	180.2	195.1	Are 81b
RSC[a)]	2.85		1.12	0.37		0.9952	3.79			194.0	Are 83
SSC (DTR 75)	2.84		1.35				4.19				Mat 82
SSC[b)]	2.61	0.52	0.67				3.80				Mat 82
BG (BrG 72)	2.67	0.64	1.08	0.41	0.12		4.38	-23.67	302.8	329.9	Are 83
Paris (LaL 80)	2.65	0.53	1.06	0.40		0.9942	4.03	-18.39	203.3	220.3	Are 81b
V_{14} (WiS 84)	2.85		1.01	0.33		0.9946	3.65	-23.67	304.2	326.8	
V_{28} (WiS 84)	2.84	0.36	1.20	0.41	0.09	0.9966	4.54	-23.75	304.5	332.8	
Bonn (MaH 89)	2.63	0.58	0.92	0.49	0.13	0.9928	4.00	-23.69	307.3	332.3	

a) $\Lambda_\pi = 0.8$ GeV, b) $\Lambda_\pi = 1.2$ GeV, $\Lambda_\rho = 1.5$ GeV.

the discrepancy. It is interesting to note that the explicit formal expression for the π-MEC contribution to the $M1$ transition including the pair term had already been given as early as 1947 by Villars (Vil 47) and for the Δ contribution in 1960 by Pearlstein and Klein (PeK 60). Since in RiB 72 the Reid hard-core potential (Rei 68) had been used, which predicts a singlet scattering length of $a_s = -16.7$ fm, considerably smaller than the experimental value, it was an open question how model dependent this result was. Subsequently it has been verified by various authors (GaH 73, CoM 73, ArF 74, KaL 77, CrT 78, KeM 78) that indeed the relative π-MEC contribution is rather model-independent. In particular, its insensitivity to the short-range behaviour was demonstrated in CrT 78 by studying unitary-equivalent transformations of the Reid soft-core potential. We would furthermore like to mention that the pair current of *ps* coupling corresponds to the contribution of a small negative-energy component in the two-nucleon wave function in a relativistic description (DrG 76).

The contribution of virtual Δ excitation in the $^1S_0$-state was first estimated by Stranahan (Str 64) who found about 3 percent enhancement. This result was confirmed in a more refined treatment in the framework of explicit isobar configurations by Arenhövel, Danos, and Williams (ArD 71, ArF 74) where the non- locality due to the intermediate-isobar propagation was included. In this work the renormalization due to the presence of a $\Delta\Delta$ component in the deuteron was also considered leading to a slight reduction of about 1 percent. The largest uncertainty came from the $M1$ N-Δ-transition strength, unknown off-shell effects and short-range regularization. Further refinements later included hadronic form factors giving a slight reduction, and small positive contributions from heavier-meson exchange, mainly ρ-exchange (Are 81b, Mat 82, Are 83). Another non-negligible positive contribution stems from Δ-MEC (LeA 83, see fig. 4.2.5).

The theoretical predictions for the total n-p radiative capture cross section for thermal neutrons ($v = 2200$ m/sec) are listed in tab. 7.1.1 as obtained for various realistic N-N potentials together with the different relative contributions of π-MEC, ρ-MEC, Δ-IC, Δ-MEC and hadronic vertex from factors to the $M1$ transition moment. One notes a relative model-independent π-MEC contribution of 2.6 - 2.9 percent. The ρ-MEC is about a factor five smaller. The IC contribution is of the order of one percent but shows a larger variation. Δ-MEC give additional 0.5 percent. Of the same order but with opposite sign is the renormalization from the $\Delta\Delta$ component in the deuteron. IC_{ren} denotes the renormalization from IC in the deuteron. Altogether one finds an enhancement of the $M1$ matrix element between 3.8 to 4.5 percent. Among the potentials that have the correct singlet scattering length a_s the Bonn r-space and the Argonne V_{28} (WiS 84) predict capture cross sections close to the experimental value within less than one percent difference (see tab. 7.1.1). Thus satisfactory agreement with experiment is achieved.

In fig. 7.1.7 we show for the Bonn r-space potential the relative contributions to the total $M1$-cross section from MEC and IC with respect to the one-body part as function of the photon energy. One notes an almost linear relative increase from 8 percent at threshold to 29 percent at $E_\gamma = 7$ MeV. The major part comes from π- and ρ-MEC increasing from 6 percent at threshold to almost 20 percent at 7 MeV. The rest is distributed between Δ-IC and Δ-MEC almost evenly. The spin-orbit current and other relativistic corrections are negligible here.

The intermediate-energy region $E_\gamma = 10 - 140$ MeV is dominated by the normal contribution as is evident from fig. 7.1.8, where the various contributions relative to the normal part (N) are shown again for the Bonn potential. While explicit MEC beyond the Siegert operators and IC become more important with increasing energy leading to an increased enhancement, the relativistic spin-orbit current has the opposite effect so that up to about $E_\gamma = 100$ MeV the resulting total cross section is almost identical to the classical non-relativistic theory (N).

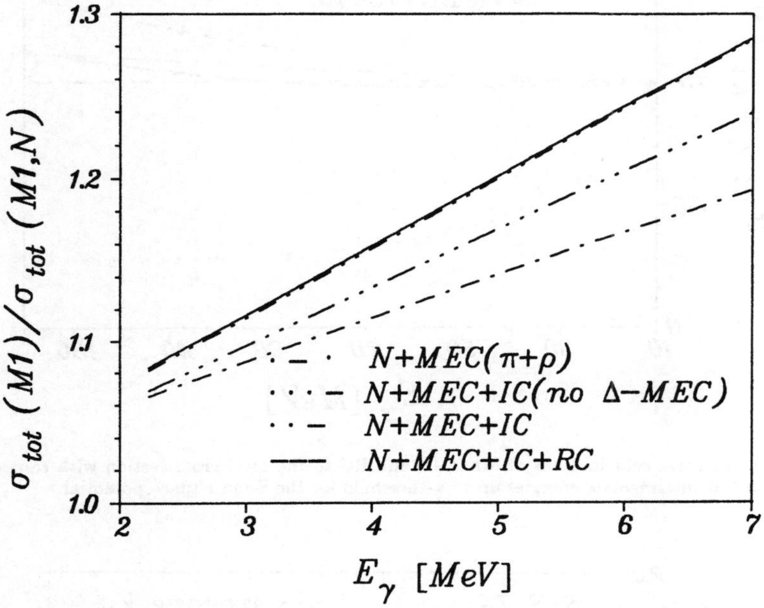

Fig. 7.1.7: Relative contributions of MEC, IC and RC to the $M1$-strength with respect to the normal $M1$ contribution at low energies.

The potential-model dependence of the total cross section and a comparison to experimental data selected according to section 6.3 is shown in fig. 7.1.9, where we have chosen the effective-range cross section of (2.18) as a reference. The theoretical uncertainties due to potential-model dependence and differences in the treatment of exchange and relativistic effects are about 4 percent at low energies and reach 20 percent at $E_\gamma = 140$ MeV. Unfortunately, the experimental errors prevent us to draw any definite conclusions as to the preference of one potential model. But there seems to be no indication of a possible breakdown of the theoretical description as claimed by Hadjimichael and Saylor (HaS 80, Are 81a).

As mentioned already above, the dominant feature between $E_\gamma = 200$ MeV and 300 MeV is the excitation of the Δ resonance, which is absent in the classical framework of nuclear physics. It can be taken into account in a diagrammatic approach (Geo 68, Wyn 71, Lag 78, Hun 80, OgK 80, AnC 81, Kan 86) or in the model of isobar configurations with explicit $N\Delta$ components in the wave functions (see section 4.2). Both methods are very similar if the IC are calculated in the impulse approximation (4.2.9). But the method of explicit IC allows a more reliable treatment of the final-state interaction when using a coupled-channel calculation for the IC (GrS 82, LeA 87). A detailed discussion of this approach can be found in LeA 87. A shortcoming is the

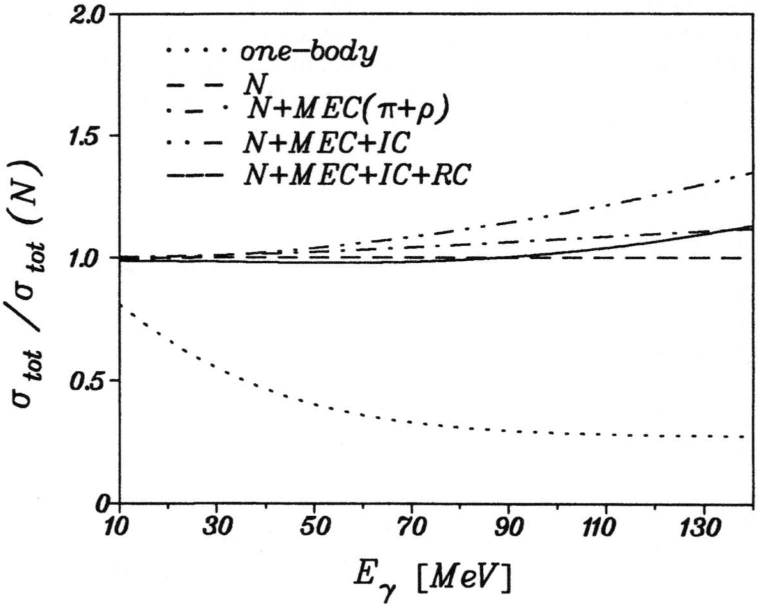

Fig. 7.1.8: Relative contributions of MEC, IC and RC to the total cross section with respect to the normal part at intermediate energies up to π-threshold for the Bonn r-space potential.

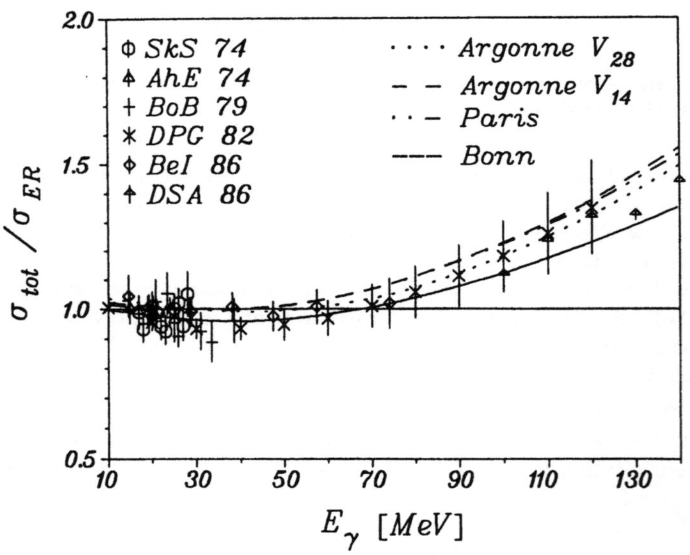

Fig. 7.1.9: Potential-model dependence of theoretical cross section and comparison to experimental data where the effective-range cross section of (2.18) is taken as reference.

neglect of explicit πNN channels which describe the pion-photoproduction processes. Their presence is only implicitly included via the Δ decay width in the complex Δ propagator. However, in a recent paper by Tanabe and Ohta (TaO 89) the missing πNN channels were considered in the unitary three-body model, which was mentioned in section 4.

A comparison between the impulse approximation (IA) and the coupled-channel method (CC) is shown in fig. 7.1.10 where the two Argonne potentials (WiS 84) have been used, i.e., V_{14} without Δ d.o.f. for IA and V_{28} with Δ d.o.f. for CC. The latter has been modified in the $^1D_2$ partial wave in order to improve the phase shifts in the Δ region by using the (RSC, 5, V_1)-model of LeA 87. One notes a marked difference. For the IA the maximum is more pronounced and above the maximum the cross section remains considerably higher than for CC, which is partly due to the neglect of the imaginary part in the Δ propagator in the IA. The most important partial wave is $^1D_2$ where the $N\Delta$ component is in a relative S-state ($^5S_2(N\Delta)$). Therefore, it is not surprising that in CC the results depend sensitively on the central $N\Delta$ interaction model, as has been discussed in LeA 87. In fig. 7.1.10 the total cross section for CC without SO current is shown in addition in order to illustrate its importance also in the Δ region.

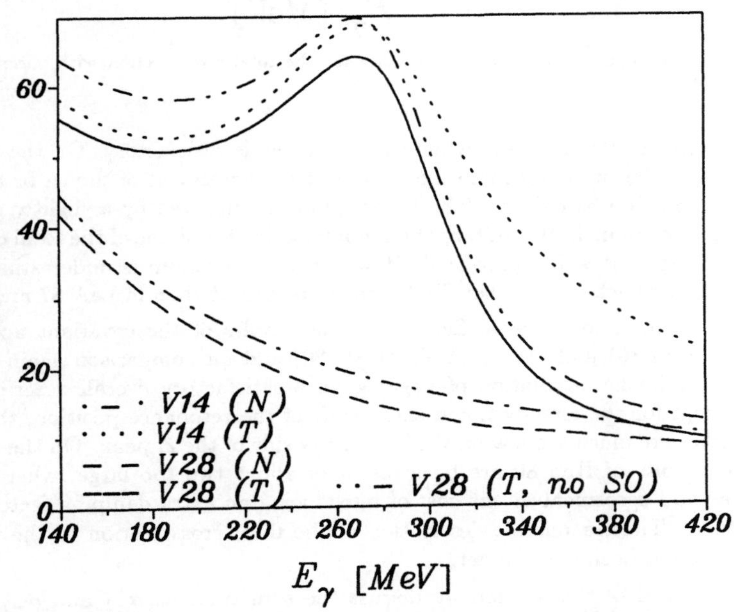

Fig. 7.1.10: Total cross section in the Δ-region for normal theory (N) and with inclusion of explicit MEC, IC, and SO (T) for the two Argonne potentials. For V_{28} the cross section without the spin-orbit current (SO) is also shown.

In fig. 7.1.11 we show a comparison of selected experimental data with results of the coupled-channel approach (LeA 87, WiL 88) with calculations of Laget (Lag 84) using the diagrammatic method and of Tanabe and Ohta (TaO 89) in the unitary three-body model. With respect to position and height of the resonance the coupled-channel approach is in satisfactory agreement with the data. On the low-energy side the RSC

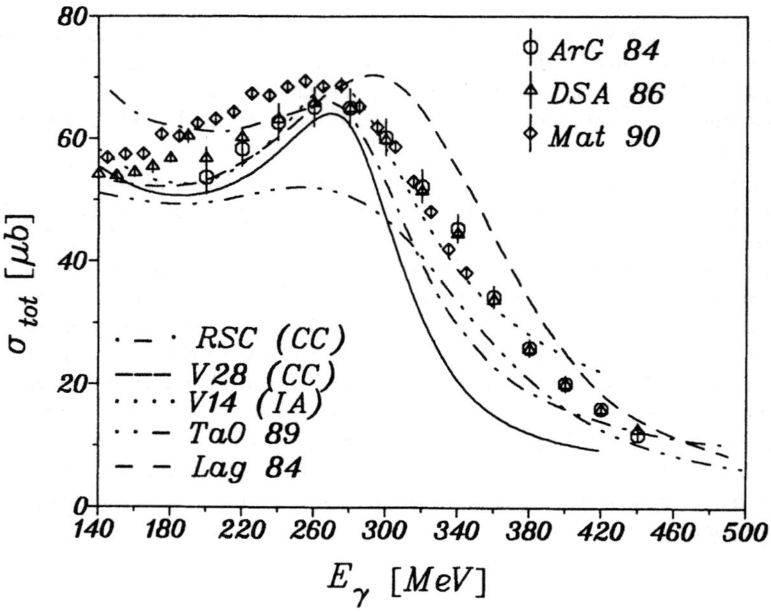

Fig. 7.1.11: Comparison of experimental data for the total cross section with various theoretical calculations.

model overestimates the data while the V_{28}-model is satisfactory. On the high-energy side both models underestimate the data. But the agreement of the IA or the result of Lag 84 is not too bad either. The latter could be improved by a slightly downshifted resonance position. In the unitary three-body model, the shape of the total cross section in the Δ region is well reproduced. However, the maximum is underestimated which can be traced back to a weaker N-Δ transition current than in LeA 87 and WiL 88.

Furthermore, we show in fig. 7.1.12 the results of the covariant approaches of Huneke (Hun 80) and Ogawa et al. (OgK 80) and for comparison again the ones of Laget. While the calculation of Laget gives a satisfactory overall description of the data except for the already noted slight shift of the resonance position, the covariant approaches are much too low on the low-energy side of the Δ peak. On the high-energy tail the results of Hun 80 are by a factor of about two too large, whereas OgK 80 achieve good agreement at the cost of introducing an extra damping factor for the Δ amplitude. Thus, a reliable description of the total cross section in the high-energy region has not been reached yet.

At the end of this section we discuss the sum rules σ_0, σ_{-1} and σ_{-2}, which are defined by

$$\sigma_n = \int_\epsilon^\infty dE E^n \sigma_{tot}(E) \, . \tag{7.1.2}$$

The integrated total cross section σ_0 can be related to the ground-state expectation value of the double commutator of the Hamiltonian with the dipole operator under the assumption that σ_{tot} can be described by pure $E1$ absorption in the long-wave-length limit.

$$\sigma_0(E1) = 2\pi^2 \langle d|[D_z,[H,D_z]]|d\rangle = \frac{\pi^2 e^2}{M}(1+\kappa) \, , \tag{7.1.3}$$

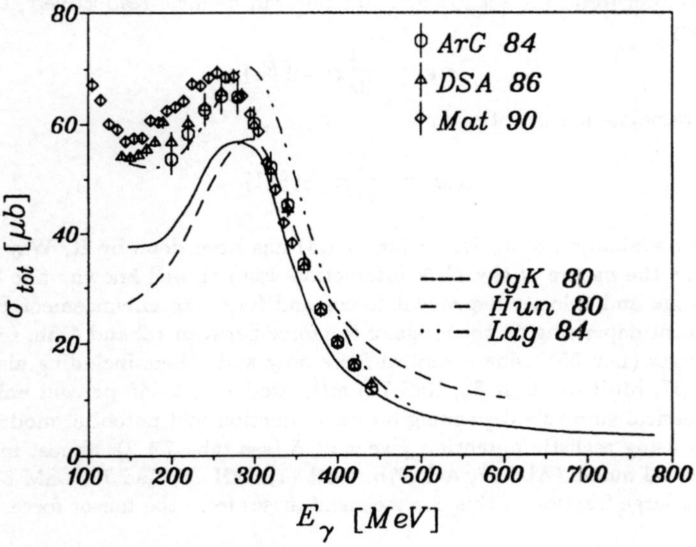

Fig. 7.1.12: Comparison of experimental data for the total cross section with various theoretical calculations using the diagrammatic method.

where

$$\kappa = \frac{2M}{e^2}\langle d|[D_z,[V,D_z]]|d\rangle \tag{7.1.4}$$

describes the enhancement due to exchange and/or velocity-dependent forces. This is the well-known Thomas-Reiche-Kuhn sum rule (Lev 60). As has been analyzed in detail in Are 83 the enhancement is a combined effect of the wave-function change by an exchange potential compared to an ordinary local potential and the two-body meson-exchange current. These two effects act in opposite directions. Without considering exchange currents but keeping the exchange potential the sum rule would be smaller than the classical value, whereas MEC enhances it.

The so-called "bremsstrahlung-weighted" sum rules σ_{-1} for $E1$ and $M1$ in the long-wave-length approximation can be related to simple ground-state expectation values (LeB 50, RuL 57, CaM 81, BoL 87)

$$\sigma_{-1}(E_1) = \frac{4}{3}\pi^2 e^2\langle r^2\rangle \tag{7.1.5}$$

$$\begin{aligned}\sigma_{-1}(M1) &= \frac{\pi^2 e^2}{M^2}(\langle M_z^2\rangle - \langle M_z\rangle^2) \\ &= \frac{\pi^2 e^2}{3M^2}\left[(\mu_p-\mu_n)^2 + 6P_D\left(1-\frac{3}{4}P_D\right)\left(\mu_p+\mu_n-\frac{1}{2}\right)^2\right],\end{aligned} \tag{7.1.6}$$

where $\langle r^2\rangle$ denotes the deuteron mean square p-n distance, M_z the z-component of the magnetic dipole operator, and P_D the D-wave probability. For the $M1$ sum rule MEC contributions have been neglected in deriving the last line of (7.1.6). The small isoscalar contribution of the deuteron D-wave in (7.1.6) was first derived correctly in CaM 81, where also explicit expressions for MEC contributions to $\sigma_{-1}(M1)$ are given.

Furthermore, from σ_{-2} for $E1$ and $M1$ one can deduce, respectively, the electric polarizability

$$\alpha_E = \frac{1}{2\pi^2}\sigma_{-2}(E_1) \tag{7.1.7}$$

and the paramagnetic susceptibility

$$\chi_M = \frac{1}{2\pi^2}\sigma_{-2}(M1)\,. \tag{7.1.8}$$

The first evaluation of $\sigma_0(E1)$ using (7.1.3) has been done by K. Way (Way 37). At that time the nature of the N-N interaction was not well known. She considered both exchange and velocity-dependent forces and found an enhancement between 25 and 90 percent depending on the range of the force between 1.2 and 4 fm, respectively. Later Levinger (Lev 55) using a central force only and others including also a tensor part (DaV 67, MuR 67, LuR 68, RaS 71) estimated a 30 to 45 percent enhancement over the classical sum rule depending on wave function and potential model. Modern evaluations using realistic potentials give $\kappa \approx .5$ (see tab. 7.1.2), almost independent of the potential model (AbS 77, ArF 77b, RuR 77, GaH 78, Had 79, CaM 81, GoL 82, VyR 82). A large fraction of this enhancement arises from the tensor force.

Tab. 7.1.2 Enhancement κ for the non-relativistic charge density (0) for relativistic one-body $\Delta\rho(1)$, two-body charge density $\rho(2)$ for pseudoscalar (ps) and pseudovector (pv) πN coupling and for IC.

Potential	κ_0	$\kappa_{0+\Delta\rho(1)}$	$\kappa_{0+\rho(2)}$	$\kappa_{0+\mathrm{IC}}$	ref.
HJ (HaJ 62)	0.524			0.738	ArF 77b
HJ	0.525	0.668	0.657 (ps)		CaM 85
			0.448 (pv)		
Yale (LaH 62)	0.529		0.719 (ps)		VyR 82
Paris (LaL 80)	0.518		0.760		GoL 82
Paris	0.491	0.497	0.704 (ps)		CaM 85
			0.499 (pv)		
RSC (Rei 68)	0.502			0.681	ArF 77b
RSC	0.502	0.530	0.649 (ps)		CaM 85
			0.483 (pv)		
SSC (DTS 73)	0.533	0.445	0.718 (ps)		CaM 85
			0.506 (pv)		

In addition, various other contributions to κ have been considered, e.g., IC (ArF 77b) and relativistic one-body and two-body charge contributions to the dipole operator (Had 79, CaM 81, GoL 82, VyR 82, CaM 85, PaR 85a). They are also listed in tab. 7.1.2. IC give via the $\Delta\Delta$ component in the deuteron an enhancement of about 0.2 for different potentials, while the relativistic one-body contribution shows a strong potential model dependence. Furthermore, the two-body contribution is sensitive to the type of πN coupling, i.e., pseudo-scalar or pseudo-vector coupling. In view of the general decrease of the total cross section at least up to $E_\gamma = 500$ MeV by the relativistic one-body current discussed above, the enhancement of the sum rule or at least not a decrease by this one-body charge correction except for the super-soft-core potential (SSC-B) is very surprising. However, one has to keep in mind that the sum rule samples all contributions up to infinity. In fact, as has been argued in CaM 85, the enhancement for those potentials not having a super soft core has to come from the very high-energy region not considered here.

This brings us to the question whether the dipole sum rule (7.1.3) is an experimental observable. Very often this sum rule is compared with the experimental cross section integrated over finite energies. But we would like to emphasize that this is not legitimate. The assumption of pure unretarded $E1$ cross section is warranted only at low energies, whereas experimental data at higher energies are sizeable and include both retardation effects and $M1$ and higher multipoles as well (Are 79). This is illustrated in tab. 7.1.3, where we show the explicitly energy-integrated theoretical total cross section as obtained with the Bonn potential up to an energy of $E_\gamma = 500$ MeV for the different ingredients together with the partial EL and ML ($L = 1, 2$) integrated cross sections.

Tab. 7.1.3 Up to $E_\gamma = 500$ MeV integrated theoretical cross sections for the Bonn r-space potential in units of the classical sum rule (30 MeVmb)

	total	$E1$	$M1$	$E2$	$M2$
one-body	0.979	0.890	0.067	0.009	0.011
N	1.568	1.479	0.067	0.009	0.011
+ MEC	1.639	1.487	0.129	0.009	0.011
+ IC	1.858	1.499	0.318	0.009	0.029
+ RC	1.726	1.400	0.292	0.008	0.024

Looking first at $E1$ one notes the above-mentioned reduction if only the one-body current is considered. Inclusion of MEC (third line) leads to an enhancement close to the double-commutator evaluation in tab. 7.1.2. But contributions from IC and RC are much smaller than in tab. 7.1.2, except for RC for the super-soft-core potential. Furthermore, $M1$ contributions from MEC and IC are sizeable and reach about 20 percent of the $E1$ strength. Higher multipoles contribute little, at most about 2 percent for $M2$. Thus, the total integrated cross section cannot be compared to the dipole sum rule in contrast to what is often stated and one should not consider this sum rule as an observable.

The situation is better for σ_{-1} and σ_{-2} due to the additional energy weighting, which favours the low-energy region, where the assumption of $E1$ and $M1$ in the long-wave-length limit is better justified. In fact, the agreement between the sum-rule prediction and the explicit integration is quite satisfactory as can be seen in tab. 7.1.4 for σ_{-1}. Relativistic corrections to the charge density contribute very little, at most about one percent (CaM 85, PaR 85a), also IC can safely be neglected (ArF 77b). These findings are confirmed in tab. 7.1.5 where a detailed evaluation by explicit integration for the Bonn r-space potential is shown. The Siegert-type MEC increase $\sigma_{-1}(E1)$ by about 20% over the one-body value but the additional contributions are very small.

For $M1$ the model-independent sum rule (7.1.6) of 0.237 mb agrees very well with the explicit integration in ArF 77b. MEC and IC contributions were evaluated in CaM

81 and found to enhance $\sigma_{-1}(M1)$ by about 30 percent. Again this agrees very well with the explicit evaluation in tab. 7.1.5. The contributions of higher multipoles are at least three orders of magnitude smaller and therefore one can identify $\sigma_{-1}(E1+M1)$ with the explicit integration over finite energies up to $E_\gamma = 500$ MeV.

Finally we present in tab. 7.1.6 the results for σ_{-2} from explicit integration. Again $E1$ is dominating but is little influenced by explicit MEC beyond Siegert and by IC and RC. $M1$ is enhanced by about 10% from MEC and IC, as expected from the threshold behaviour discussed before. Tab. 7.1.6 lists also the resulting electric polarizability and the magnetic susceptibility. The total σ_{-2} agrees very well with the experimental value of 0.69 ± 0.04 mb/MeV as analyzed in FrF 83. The $E2$ strength was investigated in Lip 77 by sum rule techniques assuming the long-wave-length limit for the charge quadrupole operator. A value for $\sigma_{-2}(E2) = 0.52$ μb/MeV was found which is close to an explicit evaluation of 0.49 μb/MeV including retardation effects.

Tab. 7.1.4 Bremsstrahlung-weighted σ_{-1} $(E1)$ from sum rule (7.1.5) and from explicit integration up to π-threshold in [mb].

Potential	N	from $\langle r^2 \rangle$ N $+\rho_2$	N + SO $+\rho_2$	$\int dE\sigma/E$ N	ref.
HJ (HaJ 62)	3.67			3.65	ArF 77b
HJ	3.668	3.706 (ps)	3.691		CaM 85
		3.670 (pv)	3.660		
RSC (Rei 68)	3.68			3.65	ArF 77b
	3.677	3.715 (ps)	3.700		CaM 85
		3.678 (pv)	3.668		
Paris (LaL 80)	3.733	3.769 (ps)	3.755		CaM 85
		3.678 (pv)	3.724		

Tab. 7.1.5 Bremsstrahlung-weighted sum rule σ_{-1}, explicitly integrated up to $E_\gamma = 500$ MeV for the Bonn r-space potential in [mb].

	total	E1	M1	E2	M2
one-body	3.288	3.059	0.219	0.0065	0.0026
N	3.949	3.720	0.219	0.0065	0.0026
+ MEC	3.986	3.721	0.256	0.0065	0.0027
+ IC	3.984	3.685	0.287	0.0065	0.0047
+ RC	3.942	3.649	0.281	0.0063	0.0041

Tab. 7.1.6 The sum rule σ_{-2}, explicitly integrated up to $E_\gamma = 500$ MeV for the Bonn r-space potential in [10^{-1} mb/MeV] electric polarizability α_E and magnetic susceptibility χ_M in [fm^3].

	total	$E1$	$M1$	α_E	χ_M
one-body	6.204	5.576	0.620	0.558	0.062
N	6.924	6.297	0.620	0.630	0.062
+ MEC	6.985	6.297	0.681	0.630	0.068
+ IC	6.925	6.225	0.693	0.623	0.069
+ RC	6.903	6.207	0.689	0.621	0.069

7.2 Angular distributions

Again, we first consider the theoretical results before we shall discuss the comparison with the experimental data as selected in section 6.3.

The angular distributions of the differential cross section show quite different behaviour in the different energy regions as is illustrated in fig. 7.2.1 for photon energies of 20, 60, 100 and 140 MeV. Very close to threshold the distribution is isotropic due to the dominant $M1$ transition into the ${}^1S_0$ antibound state. But soon with increasing energy $E1$ becomes dominant. In this region one can extract the ratio $a/b = 2\sigma(M1)/3\sigma(E1)$ from the ratio $d\sigma(0°)/d\sigma(90°)$ as outlined in section 6.2. The theoretical results on a/b for the Bonn r-space potential presented in table 7.2.1 are in satisfactory agreement with the data in table 6.2.1. With other realistic potentials including MEC contributions one finds similar results (SmB 87, RuP 89).

Tab. 7.2.1 Theoretical results on the ratio a/b

E_γ [MeV]	a/b
2.51	0.469
2.62	0.281
2.76	0.178

Above this threshold region one finds first an almost perfect $\sin^2 \Theta_p$ behaviour as one readily notes for $E_\gamma = 20$ MeV in fig. 7.2.1 because of the overwhelming $E1$ dominance in this energy region and the small optical anisotropy. Only at 0° and 180° small deviations occur, which are caused by the D-wave component in the deuteron wave function, by spin-dependent and non-central forces like the tensor force in the final-state interaction, by $M1$ and higher multipoles and other effects as will be discussed separately in section 7.3. Noteworthy is the considerably smaller cross section if only the one-body current is included, demonstrating the importance of MEC, in particular in the forward and backward direction, where the one-body prediction almost vanishes. All other interaction effects from isobars and relativistic corrections are negligible.

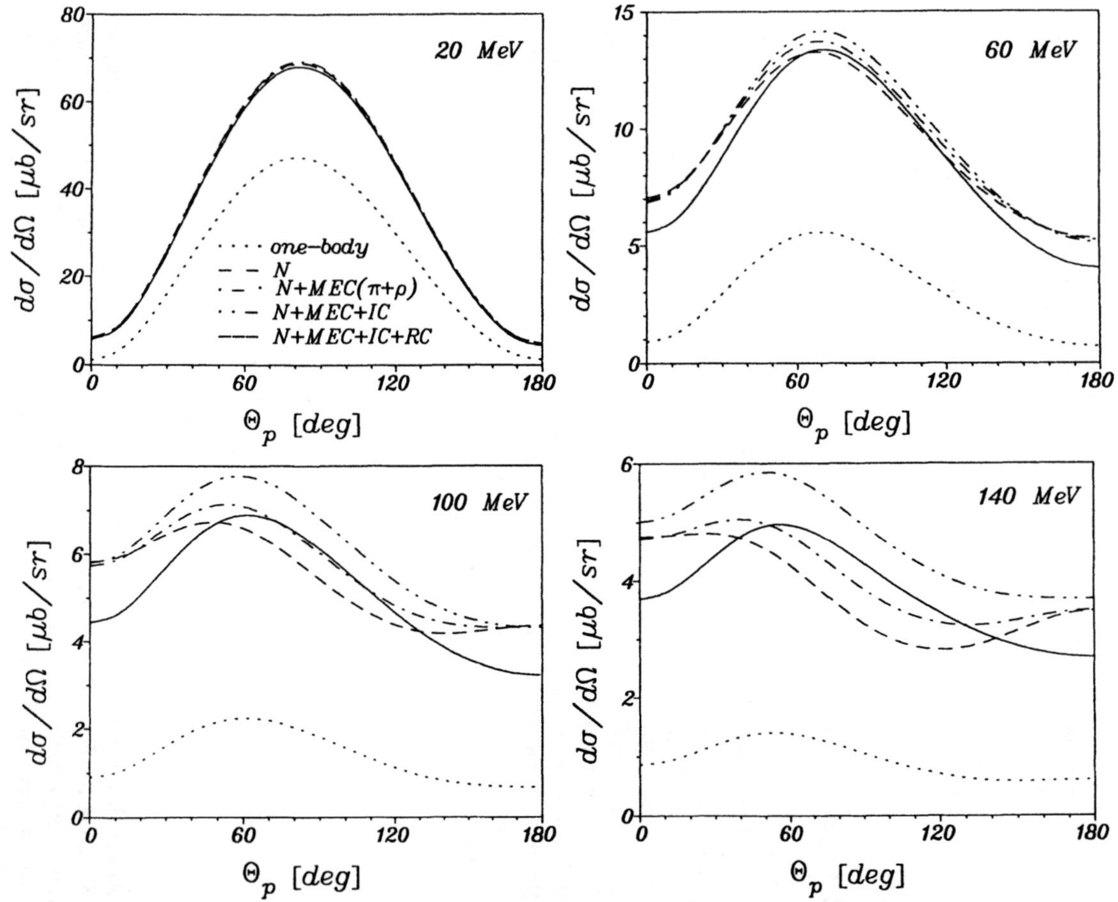

Fig. 7.2.1: Differential cross section at E_γ = 20, 60, 100, and 140 MeV calculated for the Bonn r-space potential with various contributions from MEC, IC, and RC.

This is no longer true at $E_\gamma = 60$ MeV. The ratio of the 0° cross section to the maximum is only about 0.5. The maximum is shifted from 90° to 70° and one notes a stronger forward-backward asymmetry. Around the maximum explicit MEC beyond Siegert and IC give a non-negligible increase of the cross section, whereas the spin-orbit current gives a reduction over the whole angular range, but most effective around 0° and 180°. The one-body current contributes less than 50 percent.

With increasing energy this tendency becomes more and more pronounced as the differential cross sections at $E_\gamma = 100$ MeV and 140 MeV in fig. 7.2.1 show. The maximum appears at more forward angles and the whole angular distribution becomes flatter. The contribution of the one-body current decreases to about 25 percent at E_γ = 140 MeV. Explicit MEC beyond Siegert increase the cross section sizeably between 40° and 140° and also IC by almost the same amount at 100 MeV. At $E_\gamma = 140$ MeV IC become even larger than explicit MEC. However, relativistic effects, mainly through the spin-orbit current, are of the same size but interfere again destructively leading to a sizeable decrease, in particular in the forward and backward direction.

A comparison between theoretical results using different potentials and experimental data is shown in fig. 7.2.2 for $E_\gamma = 20$ and 60 MeV. We also show the relative potential-model dependence, where the results for the Bonn r-space potential have been taken as a reference for convenience. At $E_\gamma = 20$ MeV the agreement with the data is quite satisfactory and the potential-model dependence is about 3 percent between 30° and 150° and increases to about 6 percent at 0° and 180°. This model dependence increases to about 10 percent over the whole angular range for $E_\gamma = 60$ MeV. Since the experimental uncertainties are of the same order of magnitude one may still claim good agreement within these errors.

However, the seemingly satisfactory agreement at low energies has been questioned by the recent measurements of the differential cross section ratios $\sigma(\Theta_n)/\sigma(90°)$ for neutron angles $\Theta_n = 45°, 135°$, and 155°, all lab angles and lab cross sections, at photon energies $E_\gamma = 3.5$ to 18 MeV by Stephenson et al. (StH 87). The results in fig. 7.2.3 seem to show a systematic deviation from the theory between $E_\gamma = 3.5$ to 10 MeV indicating a slight shift of the experimental differential cross section towards larger neutron angles compared to the theoretical one. As already discussed in section 6.3 this experiment was repeated by Birenbaum et al. (BiB 88) where such a deviation was not found. But the statistical uncertainty was considerably larger than in StH 87.

In any case, the systematic shift has been interpreted by Hadjimichael et al. (HaR 87) in a phenomenological analysis as a problem in the $E2$ amplitude. But they need an unreasonably large increase of the $E2$ amplitude in order to get a satisfactory fit to the data. Moreover, since the $E2$-operator is much more insensitive to the short-range part of the deuteron wave function than $E1$, it appears very difficult to invoke exotic mechanisms which affect only the $E2$ transitions and not, e.g., $E1$. In view of the importance of this question one certainly needs confirmation of the data from independent experiments.

Good agreement cannot be claimed anymore for the two higher energies of 100 MeV and 140 MeV as shown in fig. 7.2.4. First of all, the potential-model dependence increases to about 15 percent at $E_\gamma = 100$ MeV and reaches almost 25 percent at E_γ = 140 MeV. Secondly, the experimental data show also a variation of about 15 percent at 100 MeV and even more at 140 MeV. But notwithstanding these uncertainties, the average theory appears systematically higher around the maximum for $E_\gamma = 100$ MeV, and for $E_\gamma = 140$ MeV at the forward and backward directions.

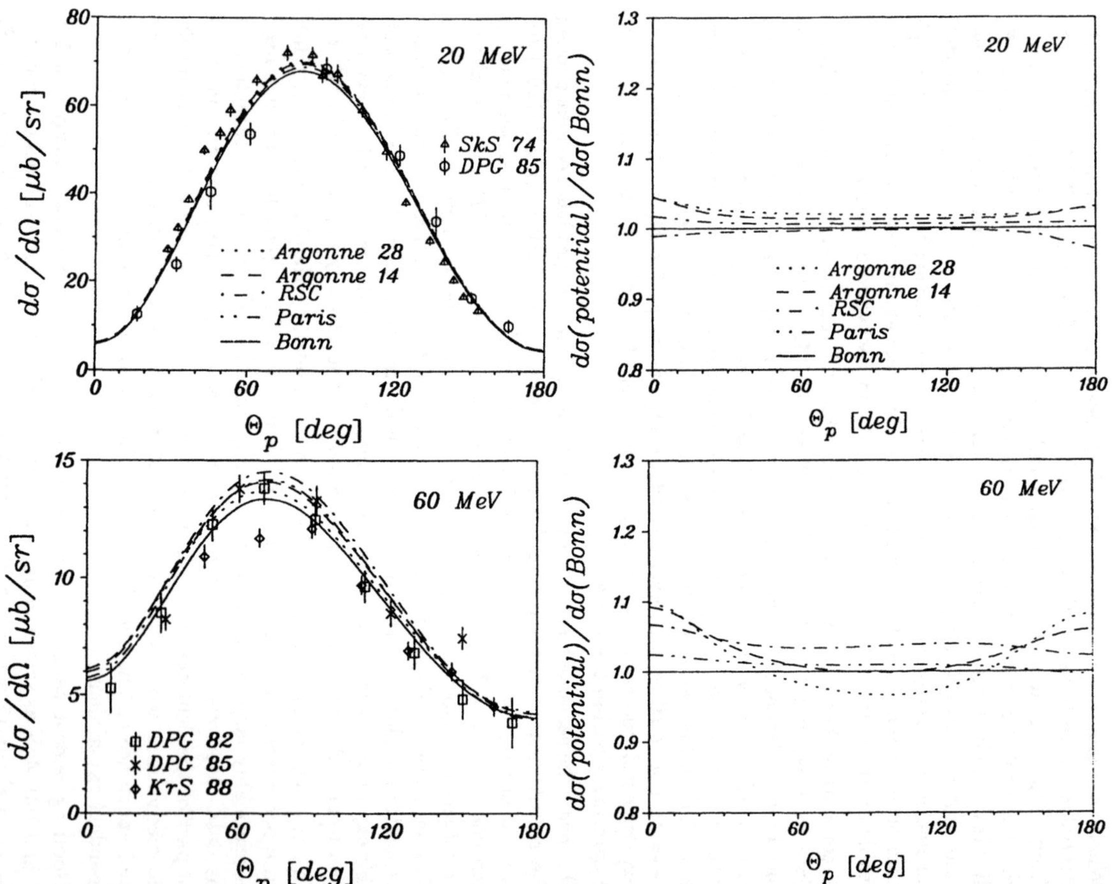

Fig. 7.2.2: Potential-model dependence of theoretical cross section including MEC, IC, and RC at $E_\gamma = 20$ and 60 MeV. Left: comparison with experimental data, right: relative contribution with respect to the Bonn r-space potential.

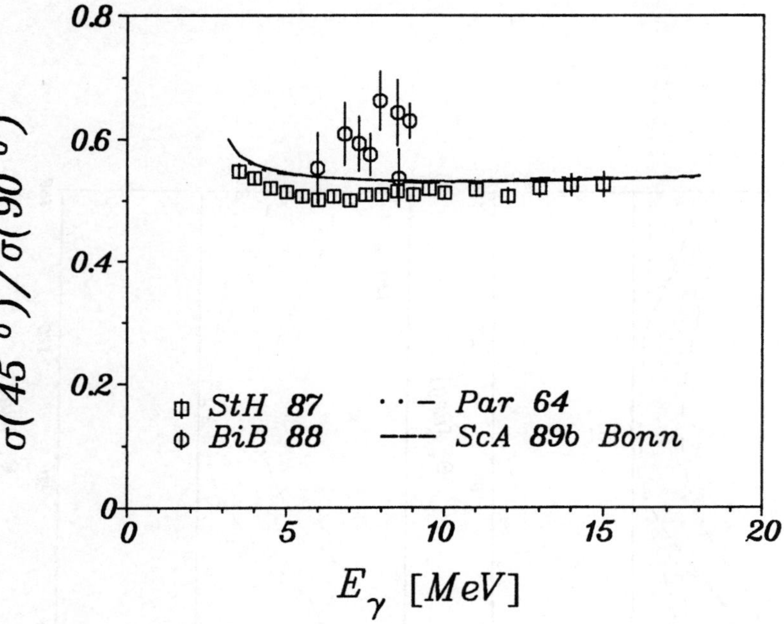

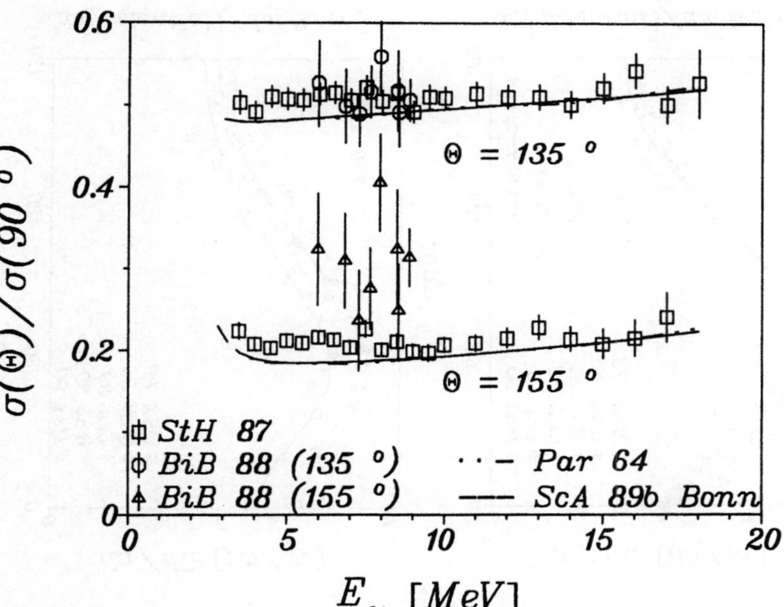

Fig. 7.2.3: Comparison between experiment and theory for the ratio of differential cross section at $\Theta_n = 45°$ (left) and $\Theta_n = 135°$ and $155°$ (right) to the one at $\Theta_n = 90°$, all lab angles.

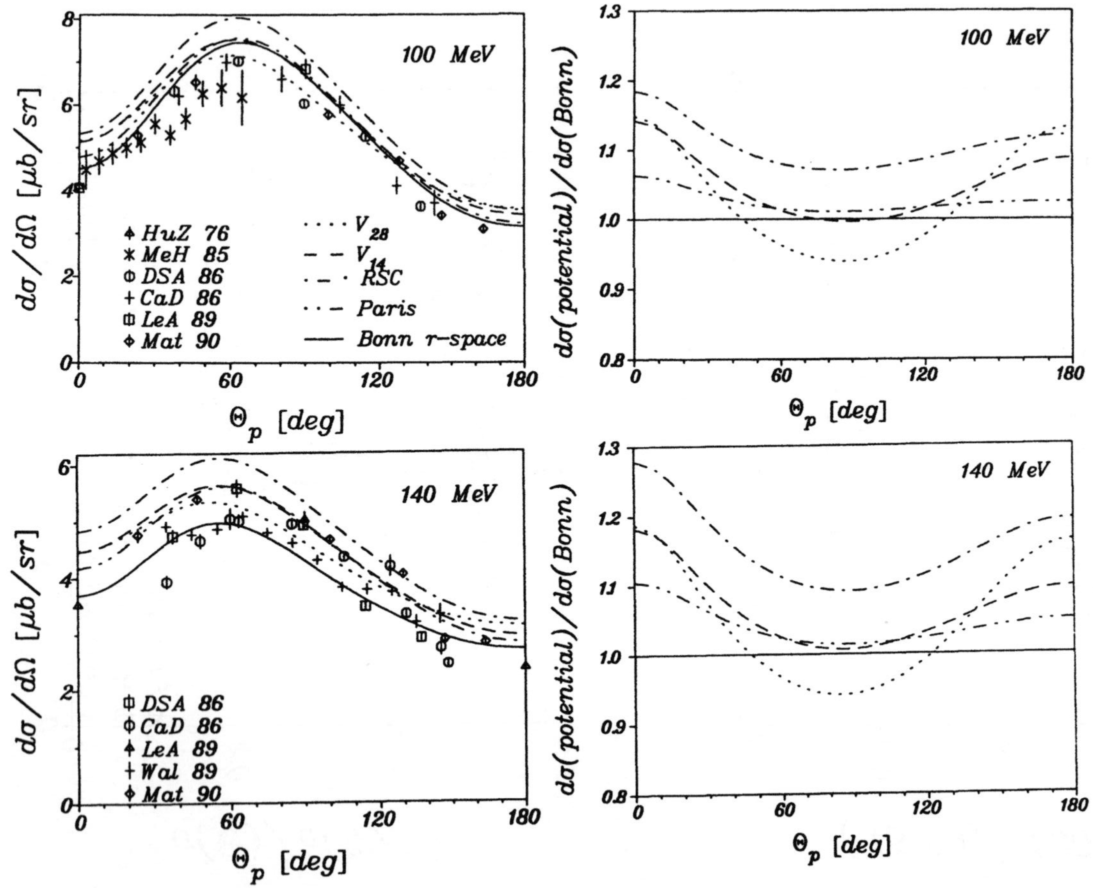

Fig. 7.2.4: As fig. 7.2.2 for E_γ = 100 and 140 MeV.

Besides the potential-model dependence the different treatments of MEC, IC, and RC introduce another uncertainty into the theoretical results. This is illustrated in fig. 7.2.5 where we show two groups of theoretical calculations of the differential cross section at $E_\gamma = 140$ MeV. One group is based on the recent Bonn r-space potential, while the other is based on the Paris potential. In the group using the Bonn potential (left part of fig. 7.2.5) the calculation of YiH 88 evaluates the MEC beyond the Siegert operator in an approximation which underestimates its contribution considerably (ScW 90). The difference between ScA 89b and WiL 88 is due to a stronger ρ-coupling in WiL 88 and additional Δ-MEC increasing the cross section while in ScA 89b relativistic corrections besides the spin-orbit current lead to a slight reduction. The differences in the calculations using the Paris potential are essentially due to differences in the IC contribution. CaM 84 use the Δ current in the static approximation neglecting the Δ propagation. The latter is included in JaW 87b and WiL 88. But in JaW 87b a weaker $\gamma N\Delta$ coupling is used and furthermore a Δ width is incorporated even below π threshold where it should be absent. Both effects result in a smaller cross section.

At higher energies due to the increasing importance of other multipoles the maximum tends to smaller, i.e., forward proton angles and becomes less and less pronounced. In other words the angular distribution becomes flatter and flatter with increasing energy and appears again almost isotropic in the Δ region around $E_\gamma = 220 - 300$ MeV. This is illustrated in fig. 7.2.6 where we show experimental data and theoretical results from WiL 88, ScA 89b, and TaO 89. Also here the spin-orbit current shows a strong influence as has been demonstrated in WiL 88. Only inclusion of it leads to a satisfactory description of the experimental data. A comparison between Δ degrees of freedom in the impulse approximation (IA) and the coupled-channel approach (CC) shows that the CC gives a slightly flatter distribution. The calculation of TaO 89 underestimates the data between $\Theta = 60°$ and 120° due to the weaker $N\Delta$ transition current mentioned in section 7.1 already. Since the unitary three-body model fails to describe the ^3P_j phase shifts of N-N scattering as remarked at the end of section 4.2, the Paris potential is used in this case and also the deuteron wave function is taken from the Paris potential.

The general variation of the differential cross section with energy is most suitably analyzed in terms of the Legendre polynomial coefficients $A_n(E_\gamma)$ as defined in (3.5.4) using again for convenience the notation A_n instead of V_{00}^n. As discussed in section 6.2 several groups (DPG 82, DSA 86, ThF 86, RoD 89b) have analyzed the various experimental data extracting experimental values for A_0 through A_3. We show in figs. 7.2.7–8 theoretical results and compare them with the experimental data already presented in figs. 6.2.3–4. In the energy range from threshold to 100 MeV (fig. 7.2.7) one finds satisfactory agreement between the experimental analysis of DPG 82 and the theory for the even coefficients A_0 and A_2, while for the odd coefficients A_1 and A_3 the theory slightly underestimates the data. MEC beyond Siegert and IC matter little except for A_2 above $E_\gamma = 40$ MeV. In the higher-energy range between $E_\gamma = 100$ and 500 MeV (fig. 7.2.8) the agreement is much less satisfactory. The theory systematically underestimates the data for A_0 up to $E_\gamma = 400$ MeV, while the data for A_1 show a larger variation around the smooth theory. Particularly large deviations occur for A_2 and A_3.

At even higher energies the various theoretical predictions for the differential cross section in fig. 7.2.9 differ quite appreciably. At $E_\gamma = 500$ MeV the IA and the diagrammatic methods of Lag 84 and AnC 81 show much too strong a forward and backward peaking which is not indicated by the data, whereas CC and the covariant methods of OgK 80 and Kan 86 lead to a reasonable description of the data. Despite this fact, none

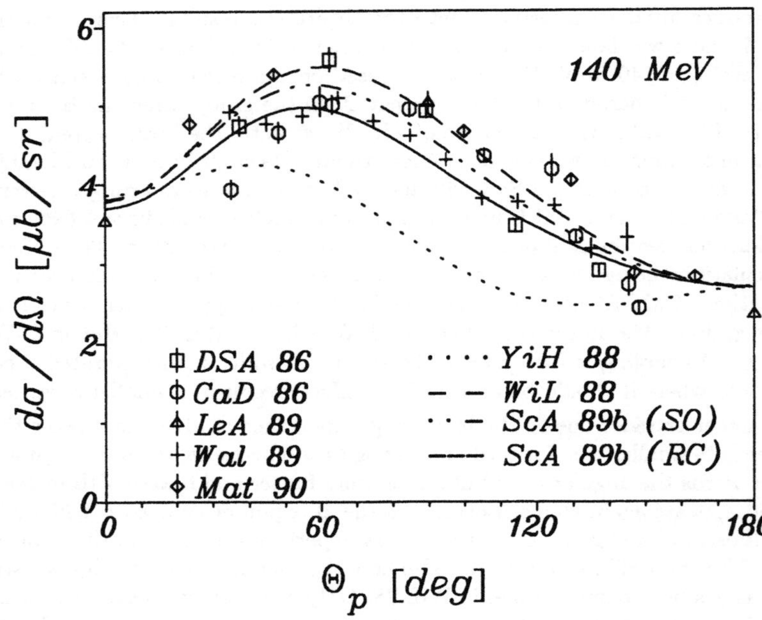

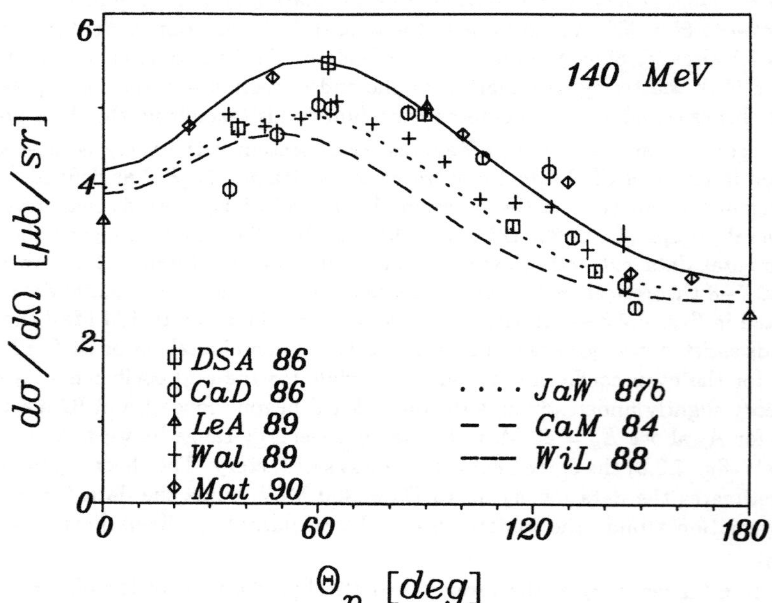

Fig. 7.2.5: Comparison of various theoretical calculations of the differential cross section at $E_\gamma = 140$ MeV using the Bonn r-space potential (left) and the Paris potential (right) with experimental data.

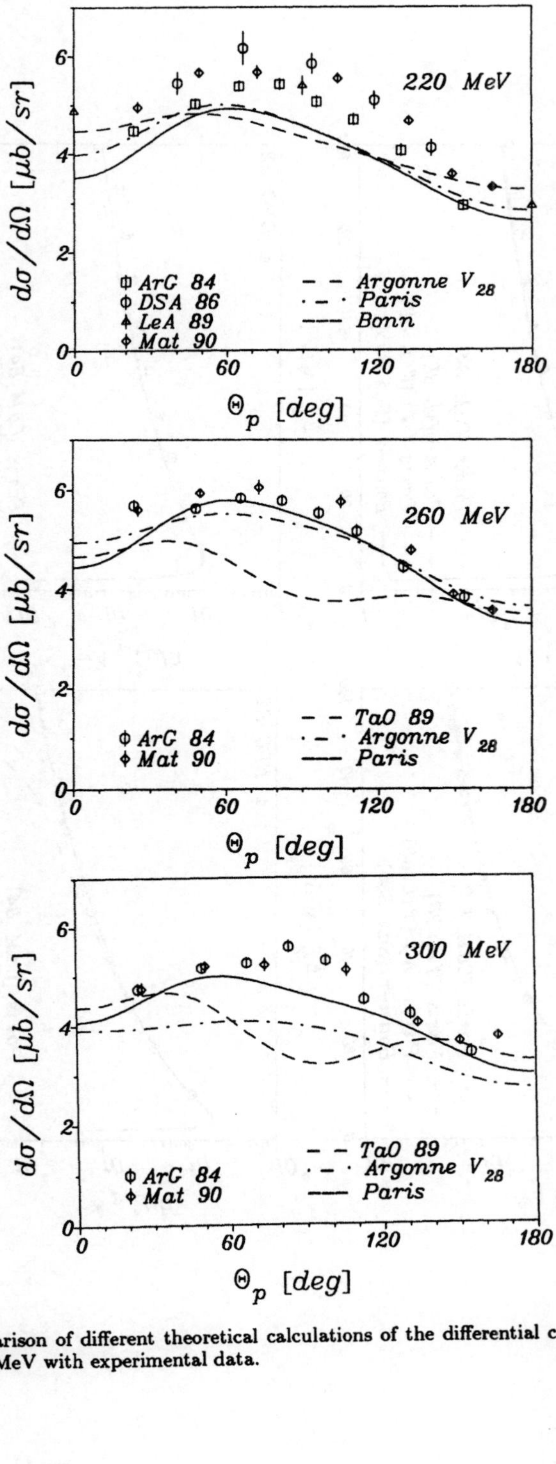

Fig. 7.2.6: Comparison of different theoretical calculations of the differential cross section at $E_\gamma =$ 220, 260, and 300 MeV with experimental data.

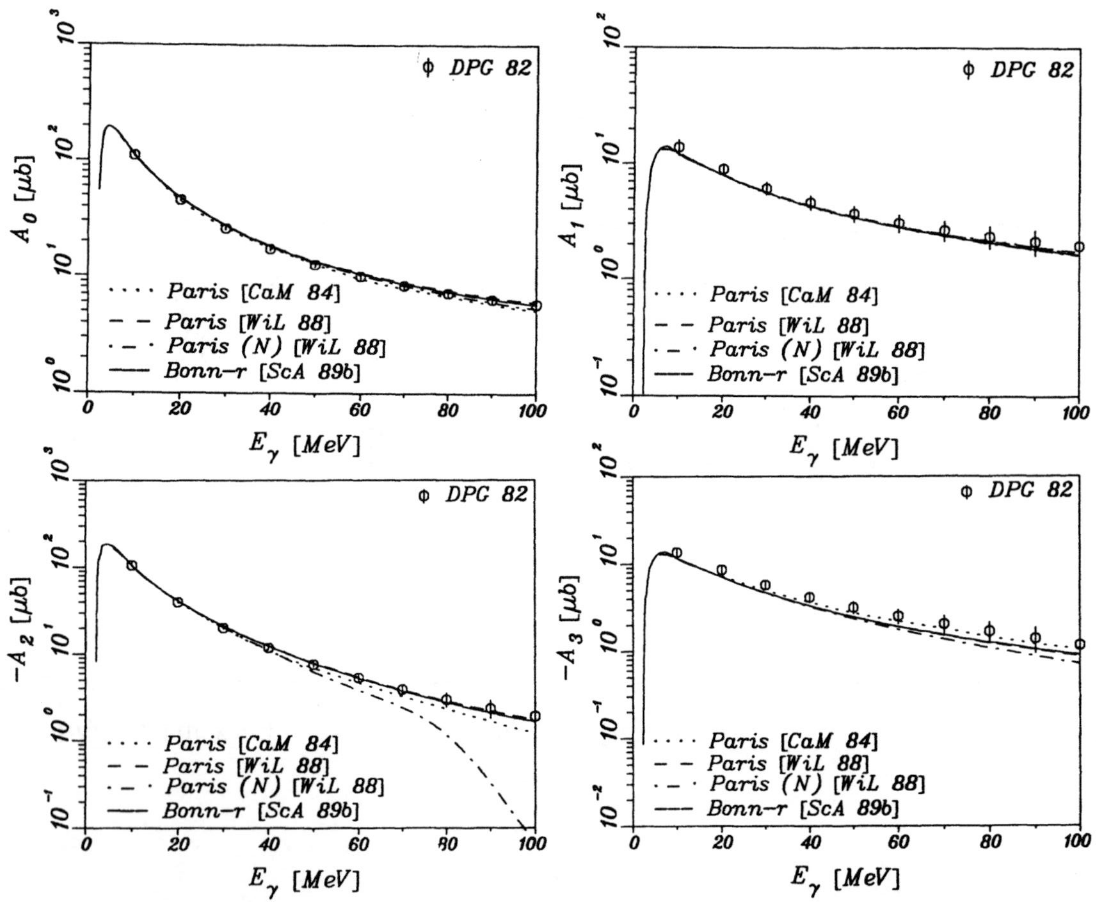

Fig. 7.2.7: Experimental and theoretical coefficients A_n $(n = 0 - 3)$ of Legendre polynomial expansion of differential cross section as function of E_γ below 100 MeV.

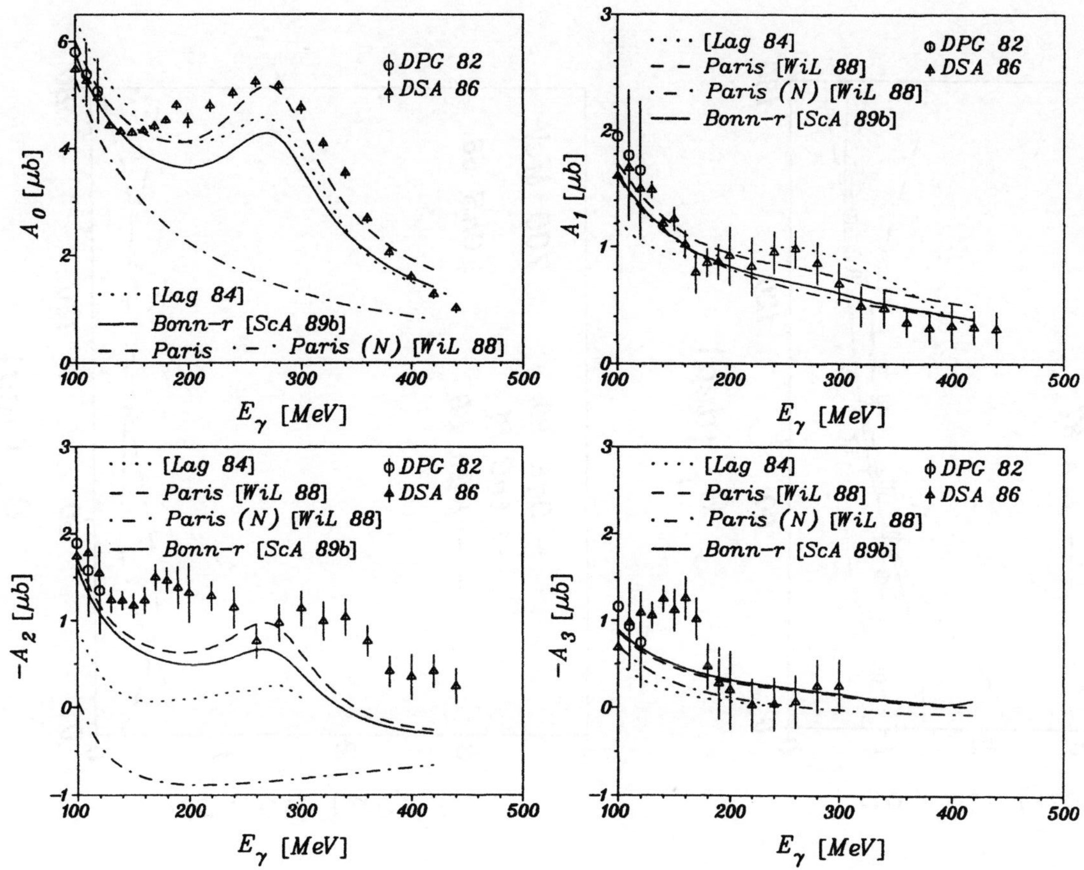

Fig. 7.2.8: As fig. 7.2.7 for photon energies between 100 and 440 MeV.

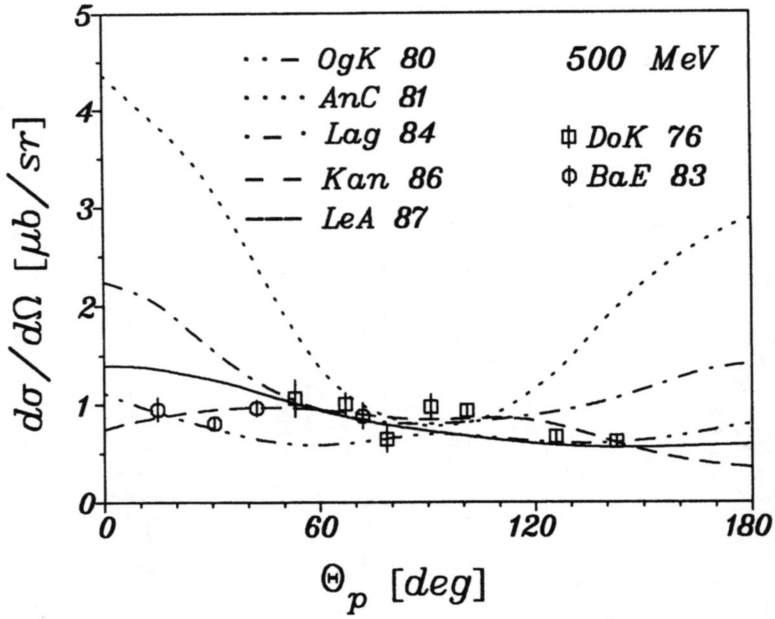

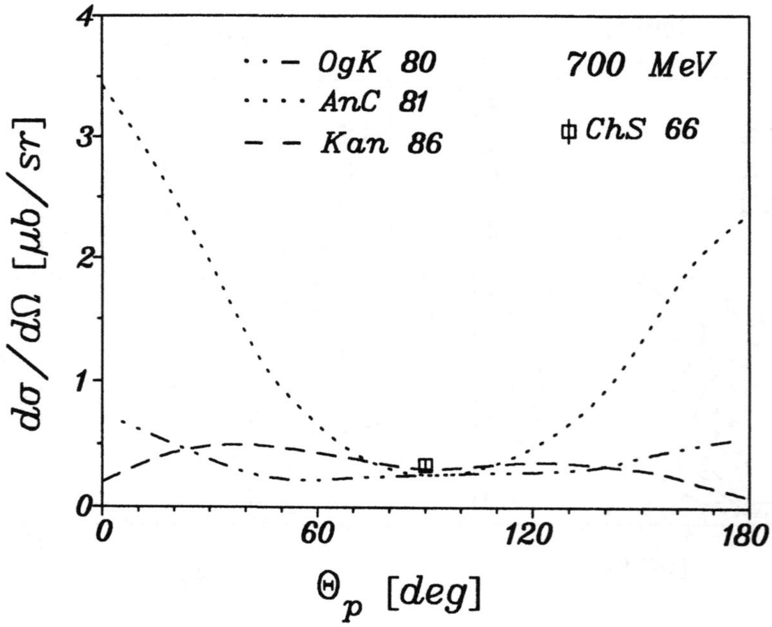

Fig. 7.2.9: Differential cross section data at E_γ = 500 and 700 MeV and various theoretical results.

of the theoretical descriptions can be considered as satisfactory. In the CC approach of LeA 87 no relativistic effects are included, which are known to be important. On the other hand, the covariant method includes final-state interaction only approximately and is poor at lower energies. Also at $E_\gamma = 700$ MeV the theoretical results are far from being satisfactory. While again the angular distribution of AnC 81 is an order of magnitude too large at forward and backward angles, the results of OgK 80 and Kan 86 show a smoother behaviour.

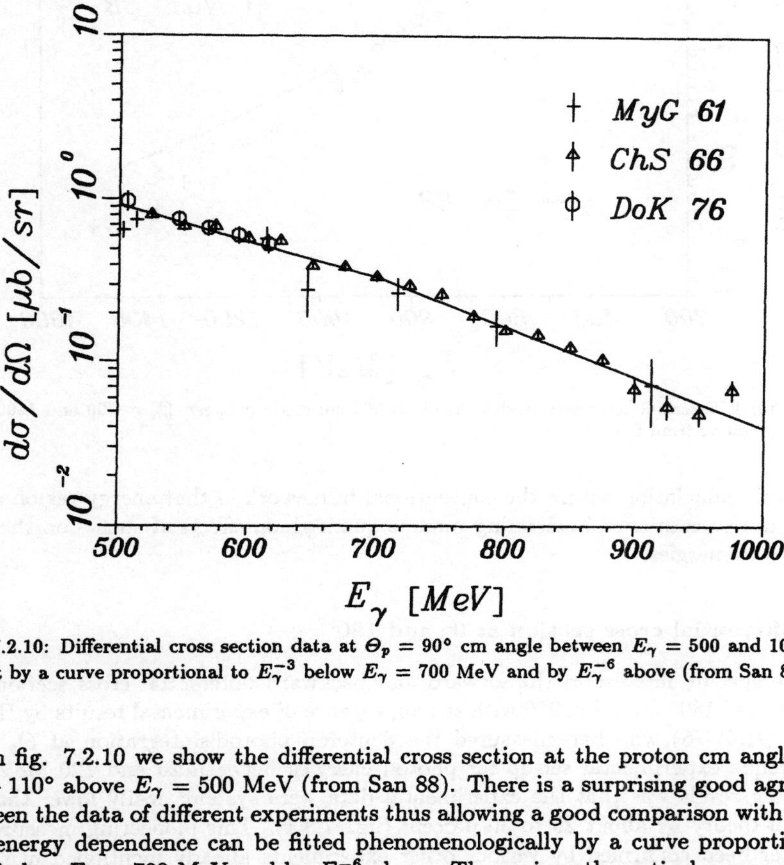

Fig. 7.2.10: Differential cross section data at $\Theta_p = 90°$ cm angle between $E_\gamma = 500$ and 1000 MeV and fit by a curve proportional to E_γ^{-3} below $E_\gamma = 700$ MeV and by E_γ^{-6} above (from San 88).

In fig. 7.2.10 we show the differential cross section at the proton cm angle $\Theta_p = 90° - 110°$ above $E_\gamma = 500$ MeV (from San 88). There is a surprising good agreement between the data of different experiments thus allowing a good comparison with theory. The energy dependence can be fitted phenomenologically by a curve proportional to E_γ^{-3} below $E_\gamma = 700$ MeV and by E_γ^{-6} above. The various theoretical curves more or less reproduce the dependence on the photon energy proportional to E_γ^{-3} in agreement with the data up to about $E_\gamma = 700$ MeV but above they fail to describe the much steeper decrease of the experimental data. This steep decrease continues at higher energies as recent new data by Napolitano et al. (NaF 88) show for $\Theta_{cm} = 90°$ at energies in the range 0.8 to 1.60 GeV (fig. 7.2.11). They could not be described by a conventional model calculation of Lee (Lee 88), which, however, can be considered only as approximate. Thus it is at present not clear whether this failure is a genuine failure of the conventional theory with nucleon, meson, and isobar d.o.f or not. On the other hand, the data of NaF 88 appear to follow a simple scaling law derived from constituent-counting relations by Brodsky and Hiller (BrH 83) assuming parton degrees of freedom for nucleons and the deuteron. In view of what we have said before, that

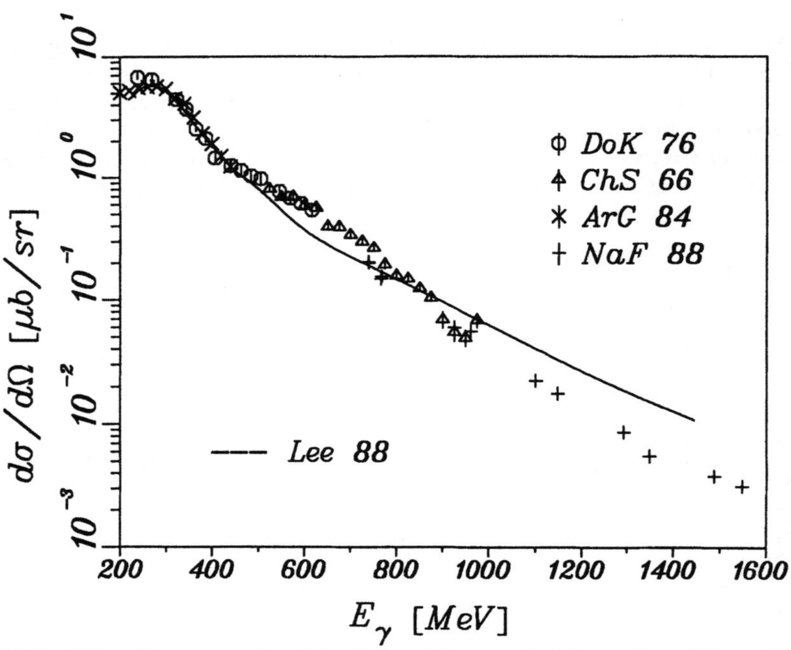

Fig. 7.2.11: Differential cross section data at $\Theta_p = 90°$ cm angle between $E_\gamma = 200$ and 1600 MeV. Theoretical curve from Lee 88.

no reliable calculation within the conventional framework in that energy region exists, one should be cautioned in claiming genuine quark-gluon effects at such – on the QCD scale – low energies.

7.3 Differential cross section at 0° and 180°

The specific interest in the forward and backward differential cross section, i.e., $\Theta_p = 0°$ and 180° arose in 1976 with the appearance of experimental results by Hughes et al. (HuZ 76) who had measured the deuteron photodisintegration at $\Theta_p = 0°$ in a unique experimental set-up for photon energies between 20 and 120 MeV. The surprising result was that the experimental data were systematically lower than the classical theory by about 20 to 30 percent (fig. 7.3.1). This pioneering measurement has later been confirmed by various other experiments already mentioned in section 6.3, some of them extending the measurements also to 180°.

The unique feature of the cross section at 0° and 180° stems from the fact that for pure $E1$ radiation – in fact $E1$ is dominating, e.g., 76 percent at $E_\gamma = 20$ MeV and 66 percent at $E_\gamma = 80$ MeV – this cross section is a sensitive indicator of the optical anisotropy of the deuteron. This can be easily understood in a classical picture. The incoming electric field $\boldsymbol{E}$ induces in the deuteron a dipole moment $\boldsymbol{P}$ and the proton is emitted with a typical dipole characteristic along $\boldsymbol{P}$. For an optically isotropic object the dipole moment is parallel to $\boldsymbol{E}$ and thus the proton emission at 0° and 180° vanishes. However, for an optically anisotropic object the induced dipole moment is tilted with respect to $\boldsymbol{E}$ and correspondingly the dipole characteristics is tilted resulting in a small non-vanishing 0°- and 180°-emission (see fig. 7.3.2).

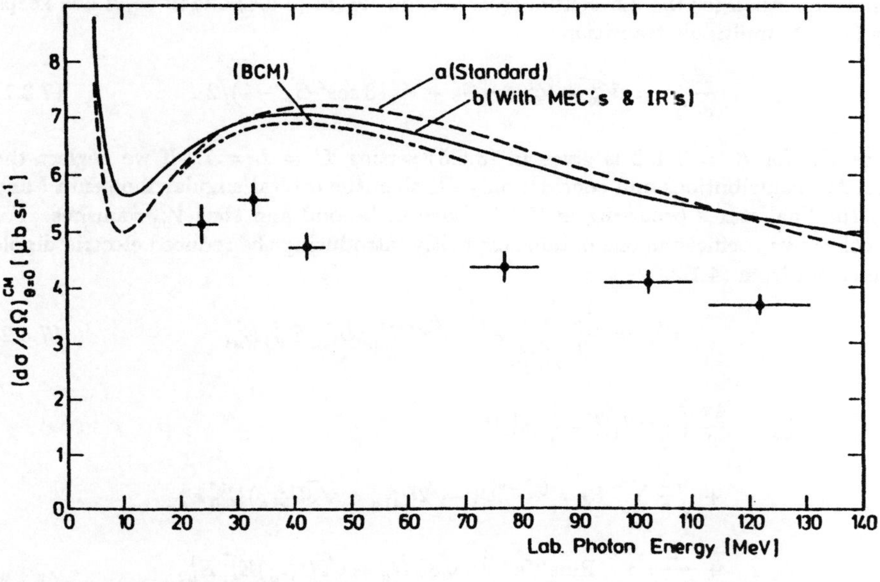

Fig. 7.3.1: Forward differential cross section data of HuZ 76 and comparison with conventional theory (from HuZ 76).

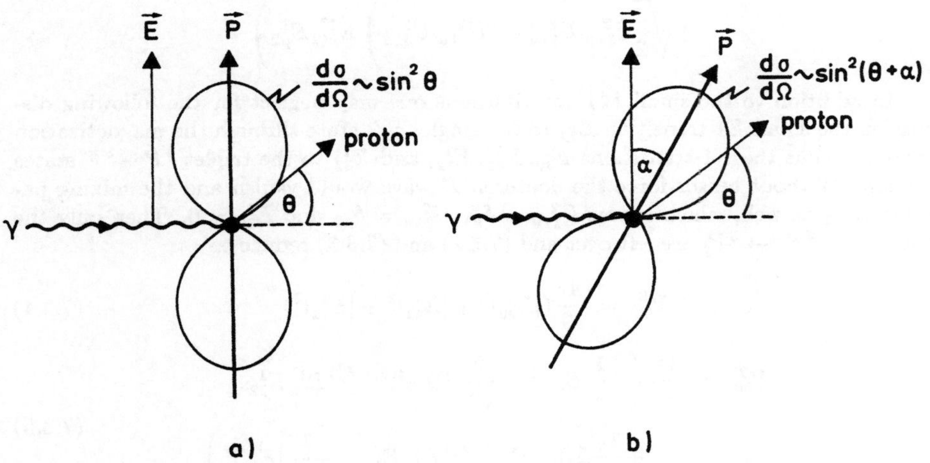

Fig. 7.3.2: Induced dipole moment for an isotropic object (a) and an anisotropic object (b) with resulting proton emission angular distribution.

Indeed, the deuteron is optically anisotropic essentially due to the tensor force of the N-N interaction, which leads to a D-state admixture in the ground state and a non-vanishing ground-state quadrupole moment. Thus this anisotropy allows also nucleon emission at 0° and 180°. In order to elaborate this in some more detail we look at the formal expression for the differential cross section. From (3.5.4) one finds, if one keeps only ($L = 1$)-multipole transitions,

$$\frac{d\sigma}{d\Omega_p} = V_{00}^0 + V_{00}^1 \cos\Theta_p + V_{00}^2 (3\cos^2\Theta_p - 1)/2 \,, \tag{7.3.1}$$

where V_{00}^K for $K = 0, 1, 2$ is given in (3.5.5) setting $L' = L = 1$. If we neglect the small $M1$ contributions and consider only $E1$ then the orbital angular momenta l and l' of the final states occurring in (3.5.5) have to be odd and thus V_{00}^1 vanishes. For the other two coefficients one obtains explicitly introducing the reduced electric dipole amplitudes from (4.1.23)

$$V_{00}^0 = \frac{4\pi}{9} \sum_{\mu'\mu j l_s} e^{i(\delta_\mu^j - \delta_{\mu'}^j)} U_{l_s\mu'}^j U_{l_s\mu}^j E_{\mu' j}^{1^*} E_{\mu j}^1 \tag{7.3.2}$$

$$\begin{aligned} V_{00}^2 = \frac{4\pi}{9} \Bigg(& \frac{1}{2}|E_{21}^1|^2 - \frac{1}{4}|E_{41}^1|^2 \\ & + \sqrt{\frac{2}{5}} \sum_\mu \text{Re}\, e^{i(\delta_\mu^2 - \delta_3^0)} (-\sqrt{2}U_{11\mu}^2 + \sqrt{3}U_{31\mu}^2) E_{30}^{1^*} E_{\mu 2}^1 \\ & + \frac{3}{2\sqrt{5}} \sum_\mu \text{Re}\, e^{i(\delta_\mu^2 - \delta_4^1)} (\sqrt{3}U_{11\mu}^2 + \sqrt{2}U_{31\mu}^2) E_{41}^{1^*} E_{\mu 2}^1 \\ & + \frac{1}{5} \sum_{\mu'\mu} \text{Re}\, e^{i(\delta_\mu^2 - \delta_{\mu'}^2)} \Bigg(-\frac{7}{4} U_{11\mu'}^2 U_{11\mu}^2 \\ & + \sqrt{\frac{3}{2}} U_{31\mu'}^2 U_{11\mu}^2 - 2U_{31\mu'}^2 U_{31\mu}^2 \Bigg) E_{\mu' 2}^{1^*} E_{\mu 2}^1 \Bigg) \,. \end{aligned} \tag{7.3.3}$$

In addition to the small $M1$ contributions one may neglect for the following discussion the small $E1$ transition E_{21}^1 to the singlet ${}^1P_1$ state through the magnetization current. Thus the $E1$ transitions $E_{30}^1, E_{41}^1, E_{12}^1$, and E_{32}^1 to the triplet ${}^3P - {}^3F$ states remain. Without tensor force the deuteron D-wave would vanish and the mixing parameter ϵ_2 as well, which means $U_{11\mu}^2 = \delta_{\mu 1}, U_{31\mu}^2 = \delta_{\mu 3}$ and $E_{32}^1 = 0$. Then only the transitions ${}^3S_1 \to {}^3P_j$ are left over and (7.3.2) and (7.3.3) reduce to

$$V_{00}^0 = \frac{4\pi}{9} (|E_{30}^1|^2 + |E_{41}^1|^2 + |E_{12}^1|^2) \tag{7.3.4}$$

$$\begin{aligned} V_{00}^2 = \frac{4\pi}{9} \Bigg(& -\frac{1}{4}|E_{41}^1|^2 - \frac{2}{\sqrt{5}} \text{Re}\, e^{i(\delta_1^2 - \delta_3^0)} E_{30}^{1^*} E_{12}^1 \\ & + \frac{3\sqrt{3}}{2\sqrt{5}} \text{Re}\, e^{i(\delta_1^2 - \delta_4^1)} E_{41}^{1^*} E_{12}^1 - \frac{7}{4\cdot 5} |E_{12}^1|^2 \Bigg) \,. \end{aligned} \tag{7.3.5}$$

Furthermore, all P-wave phase shifts will be equal if in addition the spin-orbit force is neglected. In this case the remaining reduced matrix elements $E_{\mu j}^1$ will be proportional to $(-)^j \sqrt{2j+1}$ and one then finds $V_{00}^2 = -V_{00}^0$ giving a pure $\sin^2\Theta_p$-angular distribution

$$\frac{d\sigma}{d\Omega_p} = \frac{3}{2} V_{00}^0 \sin^2 \Theta_{p'} \tag{7.3.6}$$

which corresponds to a vanishing tensor polarizability of an isotropic object (ArF 77a). This exact cancellation will obviously be destroyed by the presence of a D-wave in the deuteron and by differences between the ^3P_j phase shifts as caused by the tensor and other spin-dependent forces.

The vanishing of the differential cross section at 0° and 180° is thus not due to a suppression of matrix elements but to a destructive interference of the various $E1$-matrix elements at these angles. In view of this subtle destructive interference any small disturbance may change the $E1$-matrix elements slightly and thus may lead to a non-vanishing forward and backward cross section. It may also interfere with other small amplitudes and may show up as a relatively large effect which would go unnoticed at, e.g., 90°. Thus besides the already mentioned effects from the deuteron D-state, tensor and other non-central forces, this process is very sensitive to spin-dependent current operators, $M1$ and higher multipoles and subnuclear degrees of freedom (FrG 84).

The contribution of $M1$ and higher multipoles is shown in fig. 7.3.3, which we have taken from FrG 84. If there were only $E1$ and $M1$ one could nicely describe the data. But the other contributions cannot be neglected. In particular the relatively strong $M2$ contribution above $E_\gamma = 40$ MeV is noteworthy. We would like to point out the minimum around $E_\gamma = 10$ MeV, which is caused by the interference of the rapidly decreasing $M1$, which dominates right at threshold and the increasing $E1$ strength. Its depth and location will also be sensitive to small corrections.

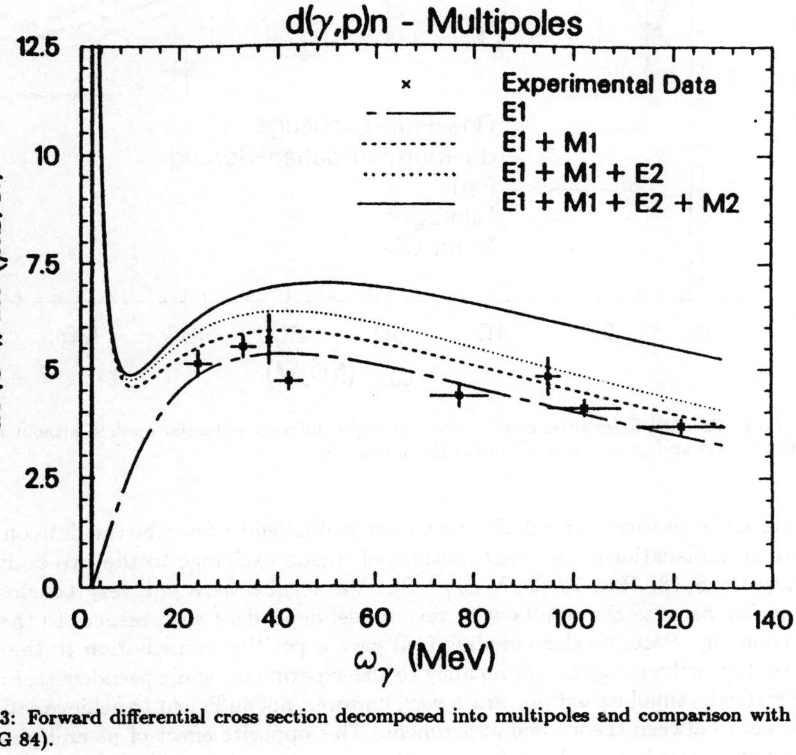

Fig. 7.3.3: Forward differential cross section decomposed into multipoles and comparison with data (from FrG 84).

The role of the D-wave probability and of the tensor force in general has been investigated by various groups (ArF 77a, Lom 77, RuS 77, CrZ 82) in the hope to close the gap between theory and experiment. But it turned out that the various realistic potentials predict 0°-cross sections which lie in a rather narrow band distinctly above the data (see fig. 7.3.4) despite the different D-wave probabilities. However, as Schulze, Saylor, and Goloskie (ScS 81) have pointed out it is not the D-wave probability which determines the 0°-cross section but rather the asymptotic normalization A_D of the D-wave. In fact, if one scales the predictions of the various realistic potentials to the experimental value of A_D then the spread of the curves in fig. 7.3.4 is even more reduced, in particular below $E_\gamma = 40$ MeV (see FrG 84).

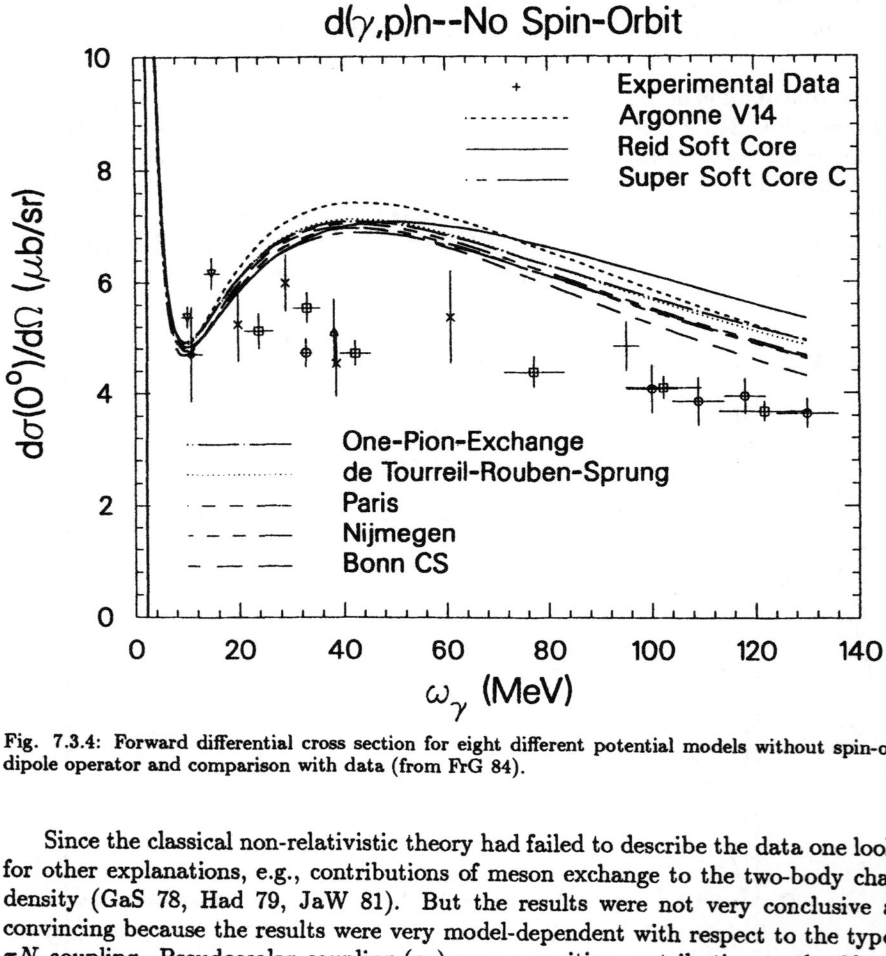

Fig. 7.3.4: Forward differential cross section for eight different potential models without spin-orbit dipole operator and comparison with data (from FrG 84).

Since the classical non-relativistic theory had failed to describe the data one looked for other explanations, e.g., contributions of meson exchange to the two-body charge density (GaS 78, Had 79, JaW 81). But the results were not very conclusive and convincing because the results were very model-dependent with respect to the type of πN coupling. Pseudoscalar coupling (ps) gave a positive contribution to the 0°-cross section thus enhancing the discrepancy to the experiment, while pseudovector coupling (pv) yielded a small reduction, which was, however, not sufficient to achieve satisfactory agreement between theory and experiment. This opposite effect of ps and pv coupling is analogous to what has been found in the sum rule σ_0 (see end of section 7.1).

Then, in 1982, Cambi, Mosconi, and Ricci (CaM 82b) pointed out that the relativistic spin-orbit contribution dominates the relativistic corrections to the one-body charge density due to the nucleon's large isovector magnetic moment (see (4.3.5)). They found that its inclusion in the dipole operator reduces the 0°-cross section by the right order almost closing the gap between theory and experiment as is shown in fig. 7.3.5, where all presently known data are presented.

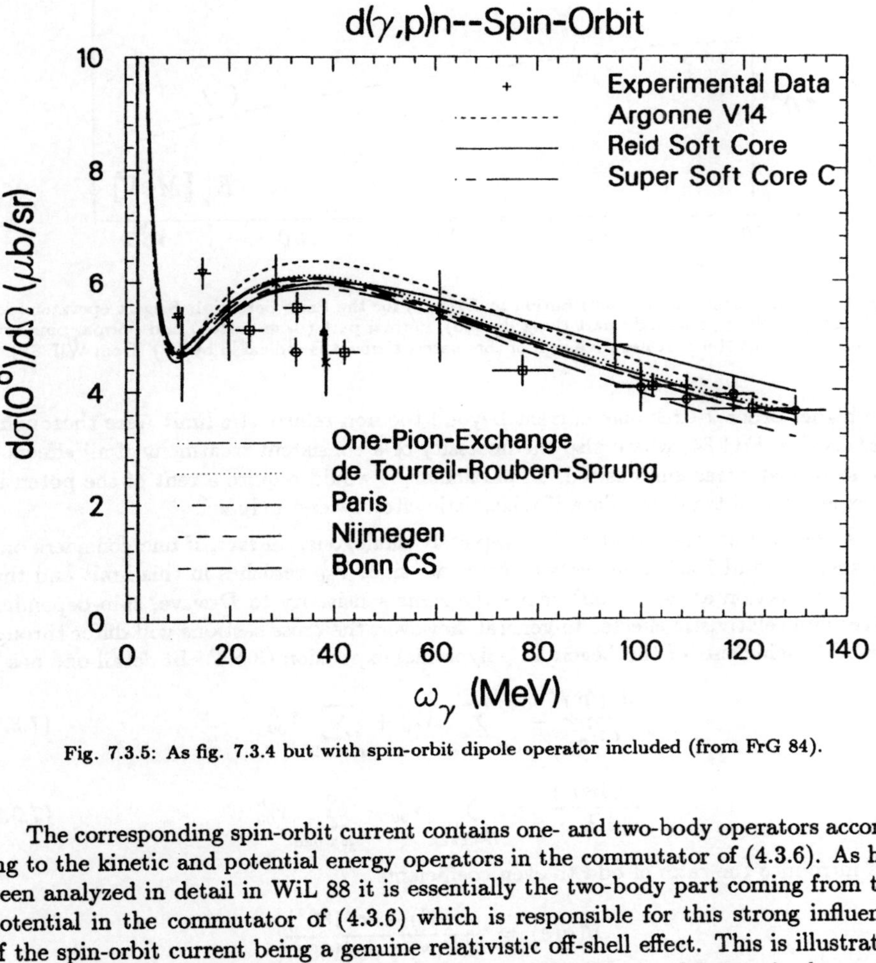

Fig. 7.3.5: As fig. 7.3.4 but with spin-orbit dipole operator included (from FrG 84).

The corresponding spin-orbit current contains one- and two-body operators according to the kinetic and potential energy operators in the commutator of (4.3.6). As has been analyzed in detail in WiL 88 it is essentially the two-body part coming from the potential in the commutator of (4.3.6) which is responsible for this strong influence of the spin-orbit current being a genuine relativistic off-shell effect. This is illustrated in fig. 7.3.6 where one of the dominant $E1$ transition-matrix elements is shown for both the normal current and the spin-orbit current as well. For the latter the one- and two-body contributions are separately shown. Clearly the two-body part dominates. One also notices that the spin-orbit current is only a rather small correction at least at low energies compared to the normal part but in view of the destructive interference at 0° and 180° its influence is relatively larger.

Even though the spin-orbit current explains the major portion of the discrepancy, other relativistic corrections cannot be neglected. The ambiguities connected with

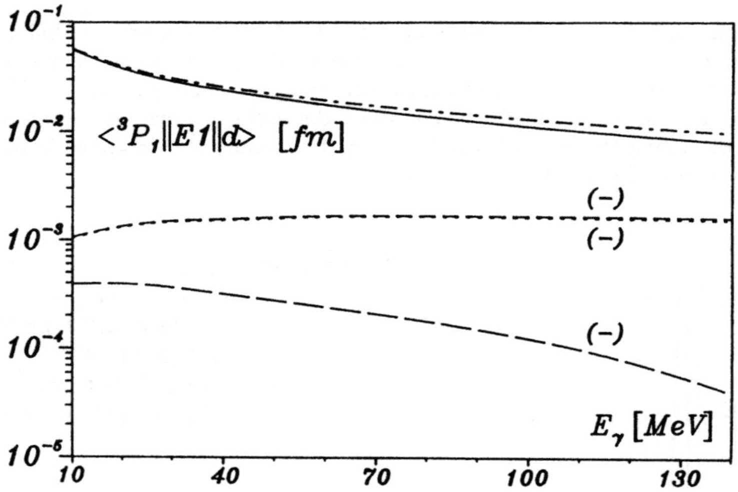

Fig. 7.3.6: Spin-orbit current contribution to $E1(^3P_1)$ for the Paris potential: Siegert operator (dotted), total (dashed), one-body part (long dashed), normal part (dash-dotted) and normal plus total spin-orbit current (full). A negative sign of the matrix element is indicated by (−) (from WiL 88).

the higher-order π-exchange current beyond the non-relativistic limit were thoroughly discussed in FrG 84, where also the necessity of a consistent treatment of all effects of the same order was emphasized. In particular, it would require a refit of the potential to experimental two-body data if relativistic effects were included.

At $\Theta_p = 180°$ the situation is completely analogous. In fact, if one considers only $E1$ then 0°- and 180°-cross sections coincide, since V_{00}^1 vanishes in this limit and thus the cross section at $\Theta_p = 180°$ shows the same sensitivity to D-wave, spin-dependent forces and relativistic effects. In general, however, the cross sections will differ through the odd coefficients of the Legendre polynomial expansion (3.5.4). In detail one has

$$\frac{d\sigma(0°)}{d\Omega_p} = \sum_{K=even} V_{00}^K + \sum_{K=odd} V_{00}^K \,, \tag{7.3.7}$$

$$\frac{d\sigma(180°)}{d\Omega_p} = \sum_{K=even} V_{00}^K - \sum_{K=odd} V_{00}^K \,. \tag{7.3.8}$$

We introduce the ratio of odd to even coefficients

$$R(o/e) = \frac{d\sigma(0°) - d\sigma(180°)}{d\sigma(0°) + d\sigma(180°)} = \frac{\sum\limits_{K=odd} V_{00}^K}{\sum\limits_{K=even} V_{00}^K} \,. \tag{7.3.9}$$

According to (3.5.5) the odd coefficients are given by interference terms of multipoles with opposite parity, while the even coefficients contain interference terms of multipoles with equal parity. Thus the deviation of the cross section ratio at $\Theta_p = 0°$ and 180°

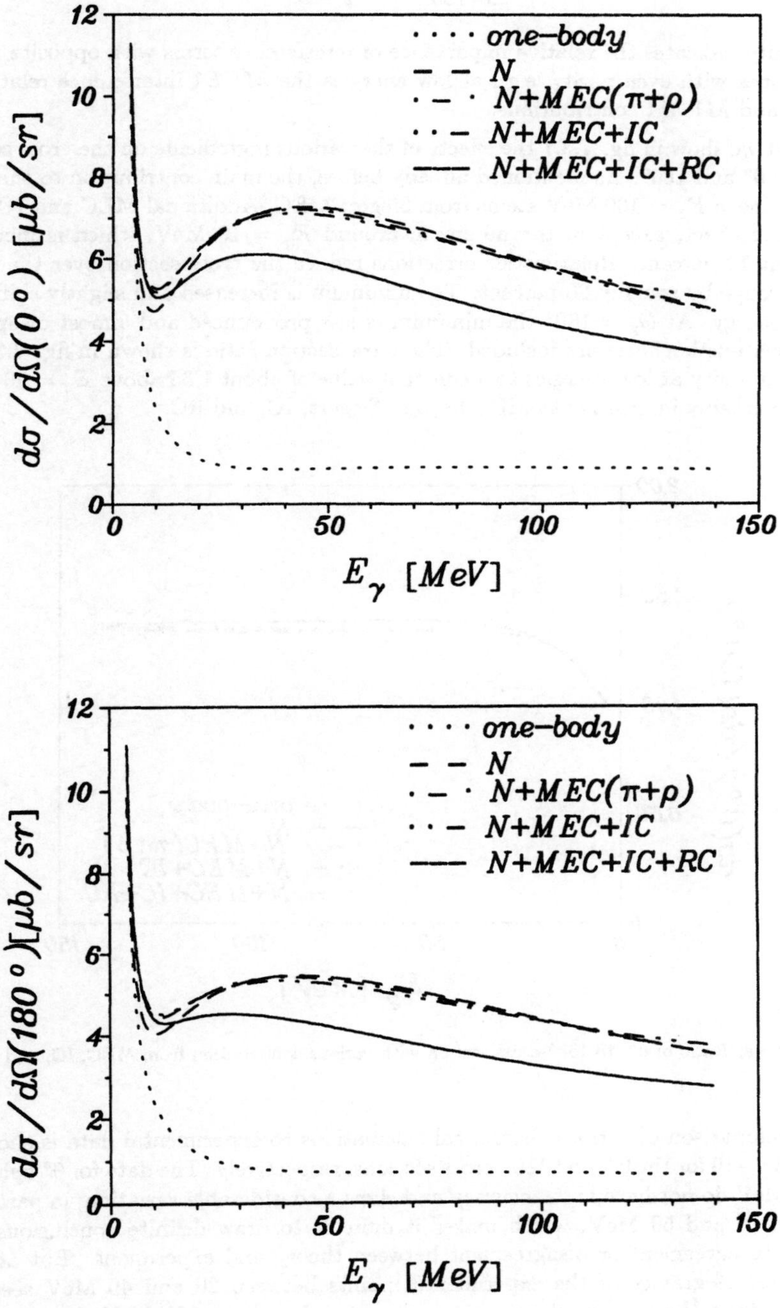

Fig. 7.3.7: Various contributions to 0°- and 180°-cross section from MEC, IC, and RC.

$$\frac{d\sigma(0^\circ)}{d\sigma(180^\circ)} = \frac{1+R}{1-R} \tag{7.3.10}$$

from unity indicates the relative importance of interference terms with opposite parity to the ones with even parity, e.g., at low energies the $M1$-$E1$ interference relative to $E1$-$E1$ and $M1$-$M1$ contributions.

First we show in fig. 7.3.7 the effects of the various ingredients on the cross section at $\Theta_p = 0^\circ$ and 180°. As mentioned already before, the main contribution to the cross section above $E_\gamma = 100$ MeV stems from Siegert MEC. Additional MEC and IC show very little effect, except in the minimum around $E_\gamma = 10$ MeV, which is increased by about 12 percent. Relativistic corrections reduce the cross section over the whole energy range by roughly 20 percent. The minimum is increased and slightly shifted to higher energy. At $\Theta_p = 180^\circ$ the minimum is less pronounced and almost disappears when relativistic effects are included. The cross section ratio is shown in fig. 7.3.8. It rises from unity at low energies to a constant value of about 1.35 above $E_\gamma = 40$ MeV. It is remarkably insensitive to MEC beyond Siegert, IC, and RC.

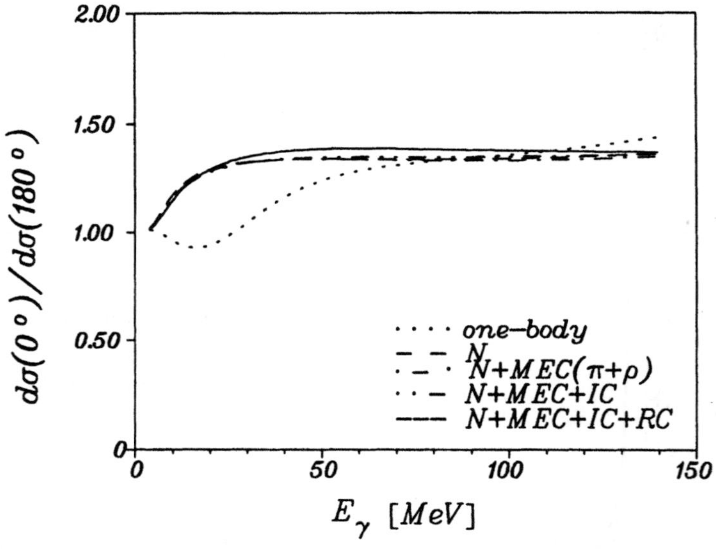

Fig. 7.3.8: Ratio of 0°- to 180°-cross section with various contribution from MEC, IC, and RC.

A comparison of various theoretical calculations to experimental data is shown in figs. 7.3.9–10 for the 0°- and 180°-cross sections, respectively. The data for 0° below E_γ = 100 MeV do not have high accuracy and show a considerable variation, in particular between 10 and 50 MeV, which makes it difficult to draw definite conclusions with respect to agreement or disagreement between theory and experiment. But at least the center of gravity of the experimental points between 20 and 40 MeV seems to be lower than the average theory. At higher energies above 100 MeV the agreement is satisfactory. The minimum around 10 MeV still awaits experimental confirmation, even though the most recent data seem to indicate this minimum (DGr 88, DGZ 89).

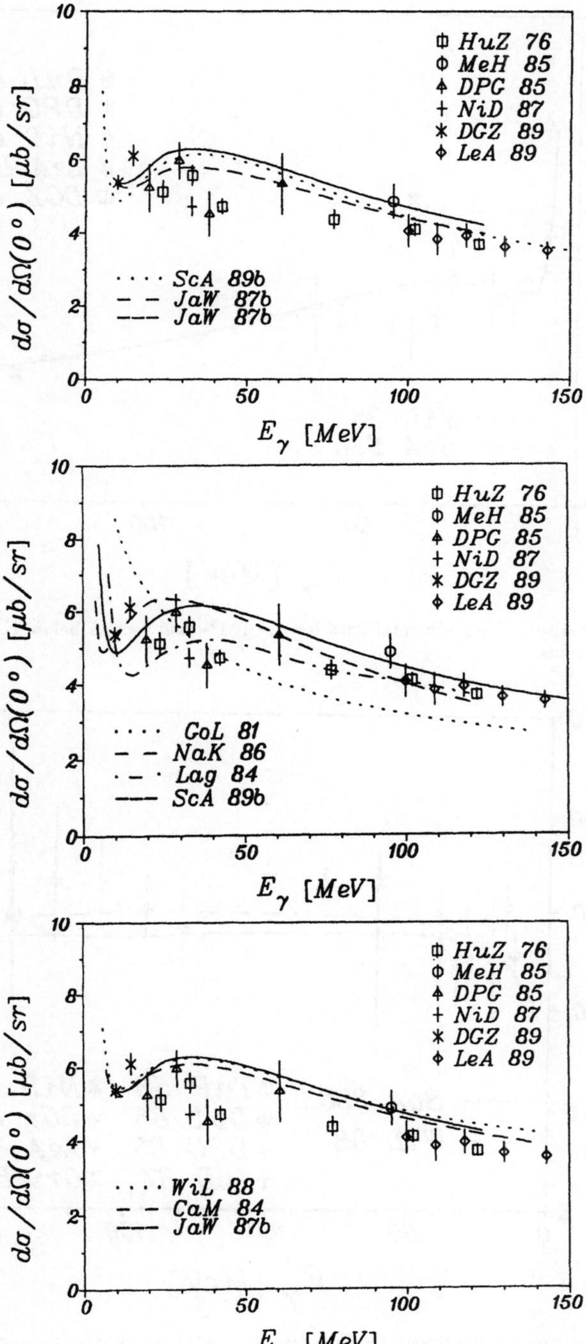

Fig. 7.3.9: Comparison of experimental data for the 0°-cross section below $E_\gamma = 150$ MeV with various theoretical calculations.

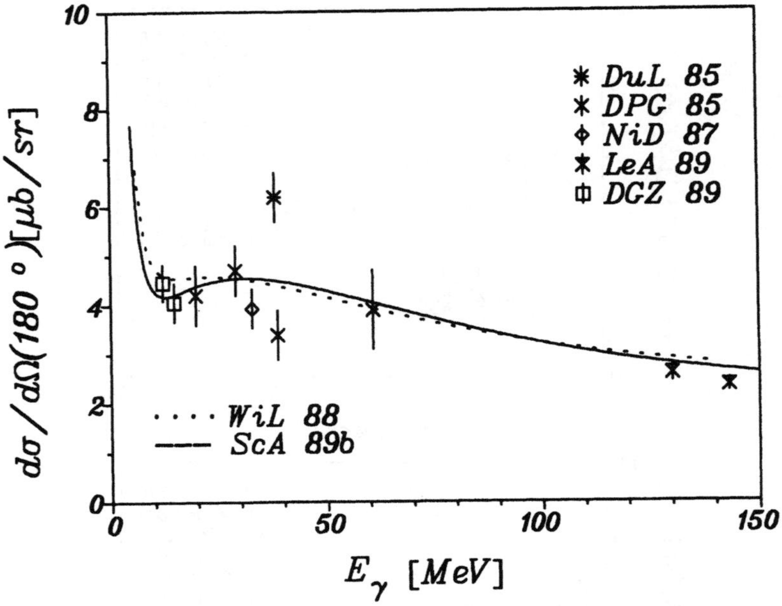

Fig. 7.3.10: Comparison of experimental data for the 180°-cross section below E_γ = 150 MeV with theoretical calculations.

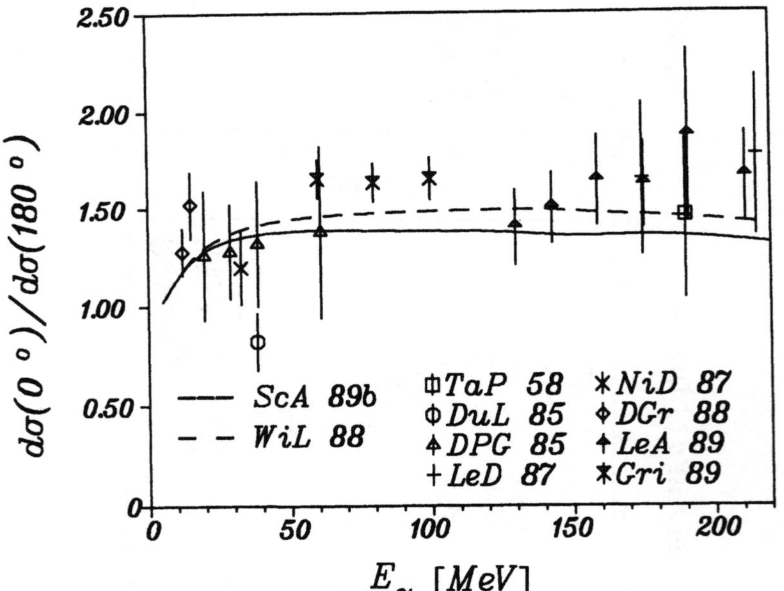

Fig. 7.3.11: Comparison of experimental data for the ratio of 0°- to 180°-cross section with theoretical calculations.

The conventional calculations of different groups show a remarkable agreement if the same potential is used (lower part of fig. 7.3.9), and, as already demonstrated in fig. 7.3.5, the potential-model dependence is very weak. Interesting is the sensitivity to isospin-symmetry-breaking effects due to differences in masses and coupling constants of neutral and charged pions as suggested in Are 81b and as shown by the dashed curve from JaW 87b in the upper part of fig. 7.3.9. Contrary to the conventional theories the diagrammatic methods (middle part of fig. 7.3.9) show a larger variation in absolute size and in the position of the minimum as well. The calculation of Lag 84 agrees nicely with the data above 40 MeV while at lower energies the cross section appears too low. On the other hand the curve of NaK 86 is too high at low energies and tends to underestimate the data above 120 MeV. The calculation of GoL 81 using the low-energy theorem fails completely.

At $\Theta_p = 180°$ the theory describes the existing data points quite well. But certainly more data are needed. The experimental 0°- to 180°-cross section ratios in fig. 7.3.11 appear on the average in agreement with the theory even though the three data points of Gri 89 lie systematically by 15 percent above the theory. It must be noted that in fig. 7.3.11 only the statistical errors are reported. In the experiment of Gri 89 a systematic error of $-(4 \pm 1)$ percent due to the asymmetry of the response of the apparatus and another systematic error of -3 percent due to background effects have been estimated.

Finally, the ratio of the differential cross section at $\Theta_p = 0°$ to the one at $\Theta_p = 90°$ cm angle is shown in fig. 7.3.12. The theoretical predictions using the Paris and Bonn potentials are almost identical and in good agreement with experimental data. Only between $E_\gamma = 100$ and 150 MeV the theory appears systematically above the data. In any case also for these ratios more precise data are needed before a definite conclusion can be reached.

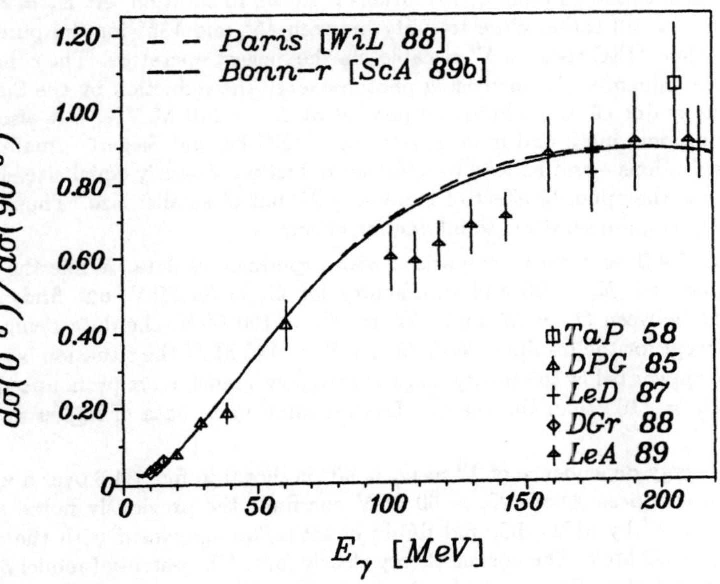

Fig. 7.3.12: Comparison of experimental data for the ratio of 0°- to 90°-cross section with theoretical calculations.

7.4 Polarization observables

The virtue of studying polarization observables has been outlined in section 6.4 where the achievement of the last decade in the development of polarized beams, oriented targets and particle polarimeters has been reviewed. Correspondingly, more and more data on polarization observables will become available in the near future, in particular with the soon available cw-machines.

We would like to recall that among the large number of polarization observables (288) that exist in deuteron photodisintegration a set of only 23 independent ones is sufficient for obtaining complete information on this process. Recently, several sets of 23 observables have been constructed in ArS 90. Since no unique solution exists, it is an open question, which set would be the most convenient one. At present, however, this seems to be a more academic question, since up to now only a very small number of observables has been studied experimentally, namely, single-polarization observables. Also the theoretical studies have focussed for a long time almost exclusively on the photon asymmetry Σ^l and on the outcoming nucleon polarization component P_y. One exception is the study of deuteron orientation in ZiA 61 and TaG 72. However, the advent of oriented deuteron targets (see, e.g., Mey 85a,b) allowing photodisintegration experiments with oriented deuterons has initiated theoretical studies of a wider range of polarization observables (Hun 80, Are 85, Are 86, ArL 86, Are 88a, ArS 90).

We now start the discussion with the simplest case of single-polarization observables, i.e., photon asymmetry Σ^l for linearly polarized photons, target asymmetries T_{IM} and single-nucleon polarization $P_y(n/p)$ for an outcoming neutron or proton. Fig. 7.4.1 shows the theoretical predictions for the photon asymmetry Σ^l as function of the proton angle for several representative energies. The Bonn r-space potential is used and the effect of the different contributions is shown in addition. At $E_\gamma = 20$ MeV the asymmetry is still rather close to unity between 45° and 135° for the pure one-body current, while MEC reduces Σ^l sizeable via the Siegert operators. The other currents show little influence. For increasing photon energy the reduction by the Siegert MEC is still the major effect – almost 50 percent at $E_\gamma = 140$ MeV –, but also the other currents become more and more important. MEC beyond Siegert – mainly through $M1$ contributions – and IC lead to a further reduction of nearly equal size. Relativistic effects show the opposite effect of increasing Σ^l but of smaller size. Thus the photon asymmetry is quite sensitive to interaction effects.

In fig. 7.4.2 we show a comparison with experimental data. While the agreement is very good for $E_\gamma = 20$ and satisfactory for $E_\gamma = 60$ MeV, one finds systematic deviations between $\Theta_p = 70°$ and 110° for $E_\gamma = 100$ MeV, the data being below the theory except for the results of WiL 88. At $E_\gamma = 140$ MeV the situation is even worse. The data appear below the theory and seem to show a much more pronounced minimum around $\Theta_p = 110°$ than the theory. But certainly more data of higher accuracy are needed.

The energy dependence of Σ^l at $\Theta_p = 90°$ is shown in fig. 7.4.3 over a wider range of photon energies. Above $E_\gamma = 50$ MeV one finds the previously noted substantial influence on Σ^l by MEC, IC, and RC in quantitative agreement with the data up to about $E_\gamma = 300$ MeV. The normal theory clearly fails. The potential-model dependence cannot be neglected. The results for the Paris potential give an overall better agreement with the data in this energy region. This potential-model dependence is also born out by the results of RuV 83b where the super-soft-core potentials give a too negative Σ^l above $E_\gamma = 100$ MeV. More recently, an improved agreement was found by PaR 85b.

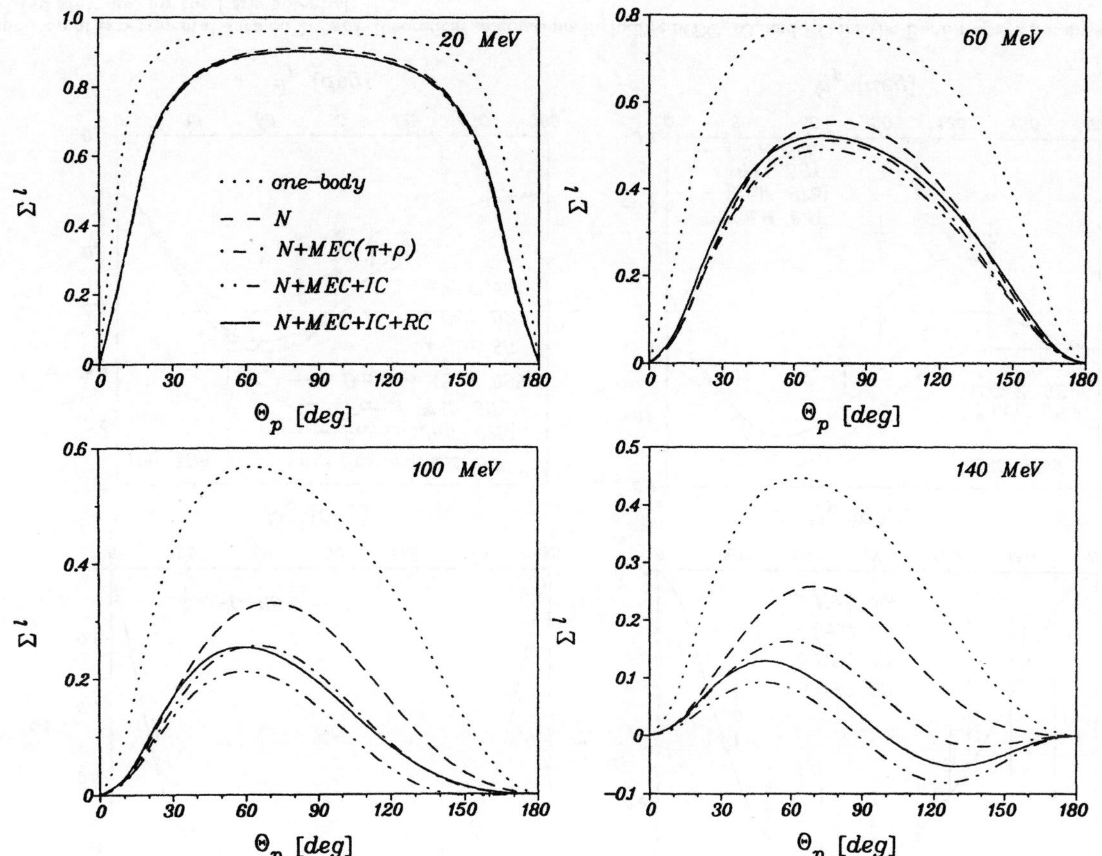

Fig. 7.4.1: Photon asymmetry Σ^l at E_γ = 20, 60, 100, and 140 MeV with various contributions from MEC, IC, and RC calculated with the Bonn r-space potential.

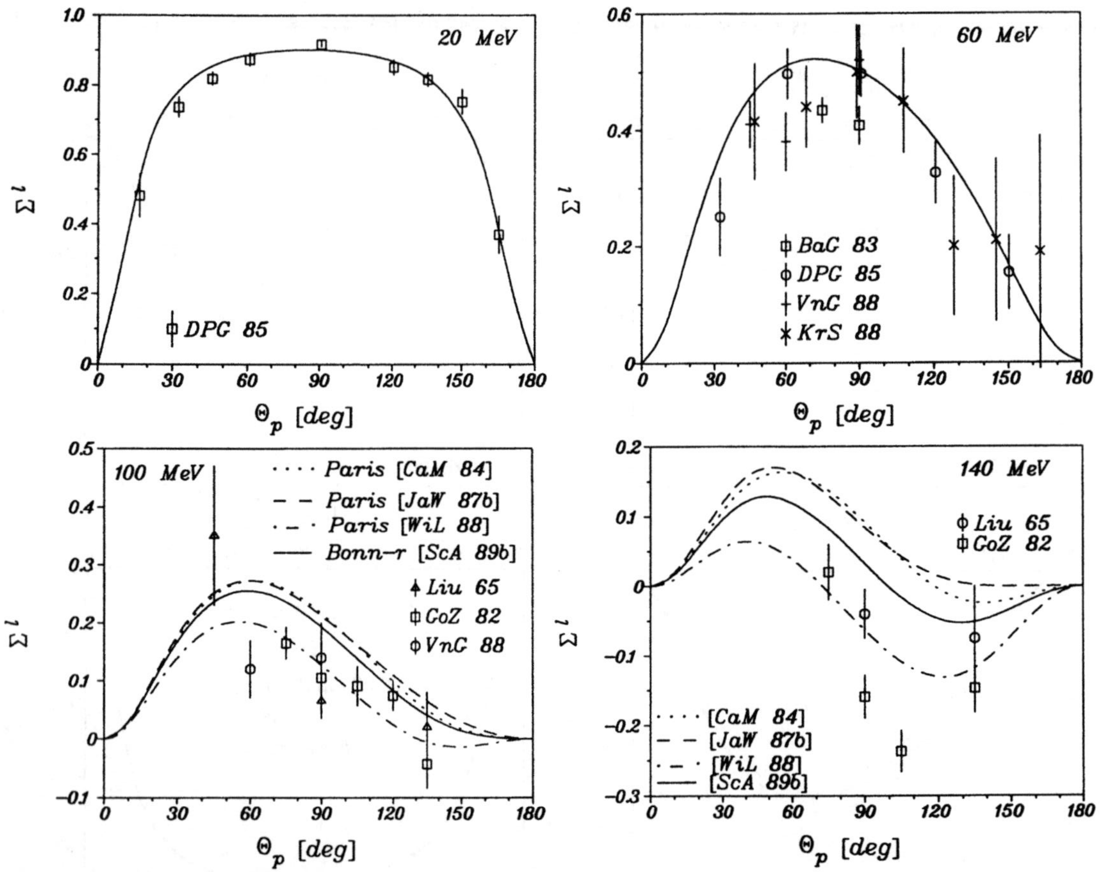

Fig. 7.4.2: Comparison of experimental data on Σ^l with theoretical calculations including MEC, IC, and RC for the Bonn r-space potential, shown in fig. 7.4.1, at E_γ = 100 and 140 MeV also for the Paris potential.

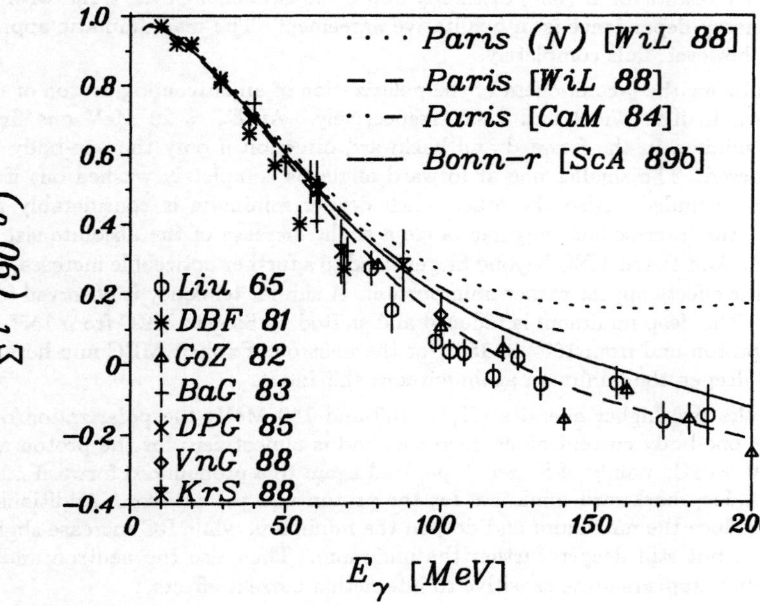

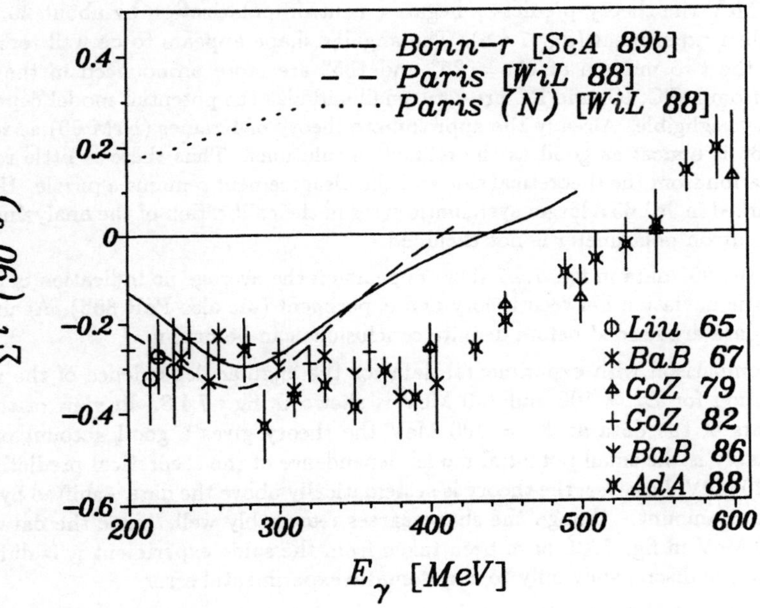

Fig. 7.4.3: Energy dependence of Σ^l at cm angle $\Theta_p = 90°$: Comparison of experimental data with theoretical calculations.

Above $E_\gamma = 300$ MeV the situation is less satisfactory. The non-relativistic calculations give results for $\Sigma^l(90°)$ distinctly above the data though the trend with respect to the energy dependence is in qualitative agreement. The diagrammatic approach of Kan 86, however, fails completely.

Results for the y-component of the polarization of an outcoming proton or neutron are shown in figs. 7.4.4 and 7.4.5, respectively. At $E_\gamma = 20$ MeV one finds two distinct minima in the forward and backward direction if only the one-body current is considered. The smaller one at forward angles is completely washed out if Siegert MEC are included. Also the other much deeper minimum is considerably reduced while for the intermediate angular range a slight increase of the absolute size can be observed. Additional MEC beyond Siegert lead to a further noticeable increase, whereas relativistic effects appear rather unimportant. A similar tendency is observed for $E_\gamma =$ 60 MeV. The deep minimum is reduced and shifted by Siegert MEC from 155° to 140° for the proton and from 165° to 150° for the neutron. Explicit MEC and here also IC and RC deepen the minimum again without shifting it.

For the two higher energies ($E_\gamma = 100$ and 140 MeV) the polarization obtained with the one-body current alone decreases and is almost zero for the proton at $E_\gamma =$ 140 MeV. MEC, mainly of Siegert type, lead again to a pronounced forward maximum and to a deep backward minimum for the proton and the neutron. Additional MEC and IC reduce the maximum and deepen the minimum, while RC increase slightly the maximum but still deepen further the minimum. Thus also the neutron and proton polarization appears quite sensitive to interaction current effects.

Experimental results on neutron polarization at low photon energies are shown in figs. 7.4.6–7 and compared to theoretical calculations. At the very low energy of E_γ = 2.75 MeV the theory predicts a negative neutron polarization by about 20 percent larger than experiment (fig. 7.4.6). The angular shape appears to be well reproduced though the two minima at $\Theta_n = 25°$ and 155° are more pronounced in the theory. Effects from MEC, IC, and RC are very small and also the potential-model dependence is almost negligible. Already the approximate theory of Kramer (KrM 60) as reported in JeJ 65 is almost as good as the refined calculations. Thus there is little room for modification from the theoretical side and the disagreement remains a puzzle. However, as remarked in JeJ 65 a larger systematic error in the calibration of the analyzing power of the neutron polarimeter is not excluded.

For the 90°-data in fig. 7.4.7 there appears on the average no indication of a major systematic deviation between theory and experiment (see also PaR 85b). Again, more precise data are needed before definite conclusions can be drawn.

A comparison with experimental data for the angular dependence of the neutron polarization for $E_\gamma = 100$ and 140 MeV is shown in fig. 7.4.8. In view of the large error bars of the data at $E_\gamma = 100$ MeV the theory gives a good account of them. Remarkably is the small potential-model dependence of the theoretical predictions. At $E_\gamma = 140$ MeV, however, the theory is systematically above the data – shifted by almost a constant amount – though the shape agrees reasonably well. Since the data at 100 and 140 MeV in fig. 7.4.9 have been taken from the same experiment it is difficult to attribute the discrepancy only to a systematic experimental error.

For the proton polarization we show angular distributions for photon energies $E_\gamma =$ 400, 450, and 500 MeV in fig. 7.4.9, which we have taken from LeA 87. The comparison of the various theoretical results clearly shows that the non-relativistic approach with explicit Δ d.o.f. gives a satisfactory description of the data provided that the N-Δ

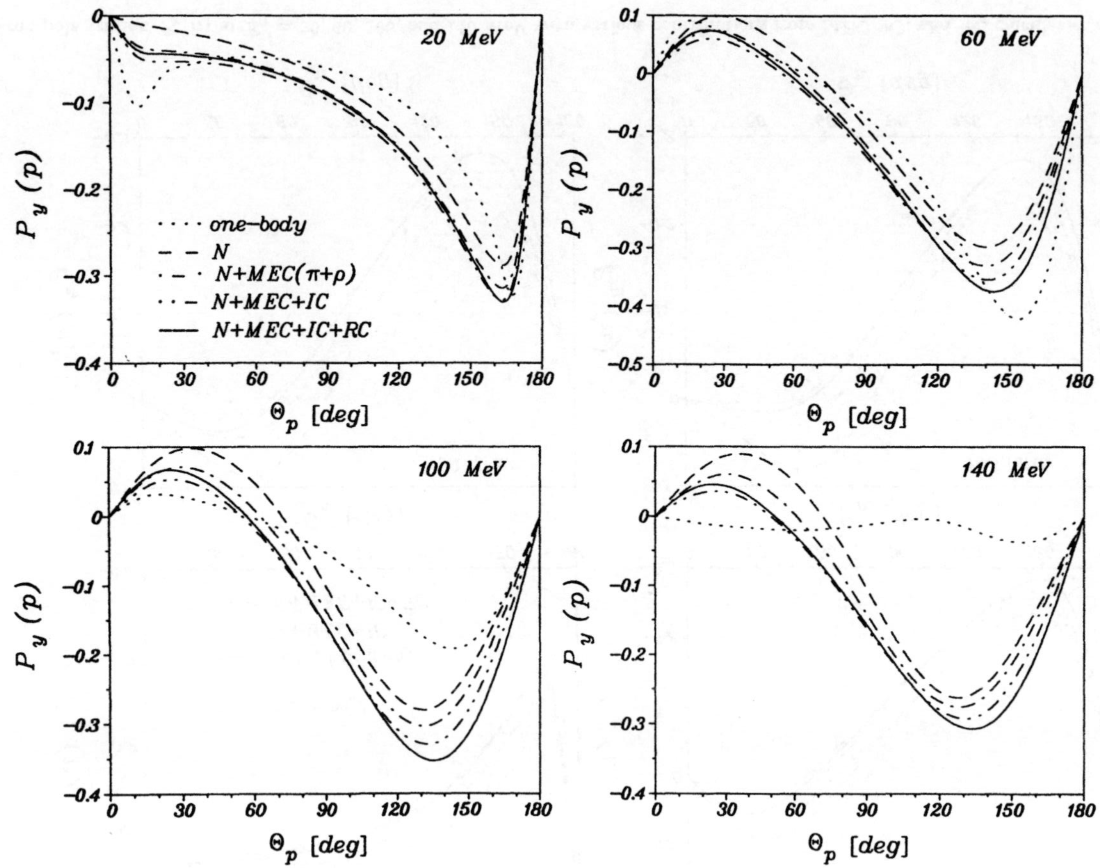

Fig. 7.4.4: Proton polarization $P_y(p)$ at $E_\gamma = 20$, 60, 100, and 140 MeV with various contributions from MEC, IC, and RC calculated with the Bonn r-space potential.

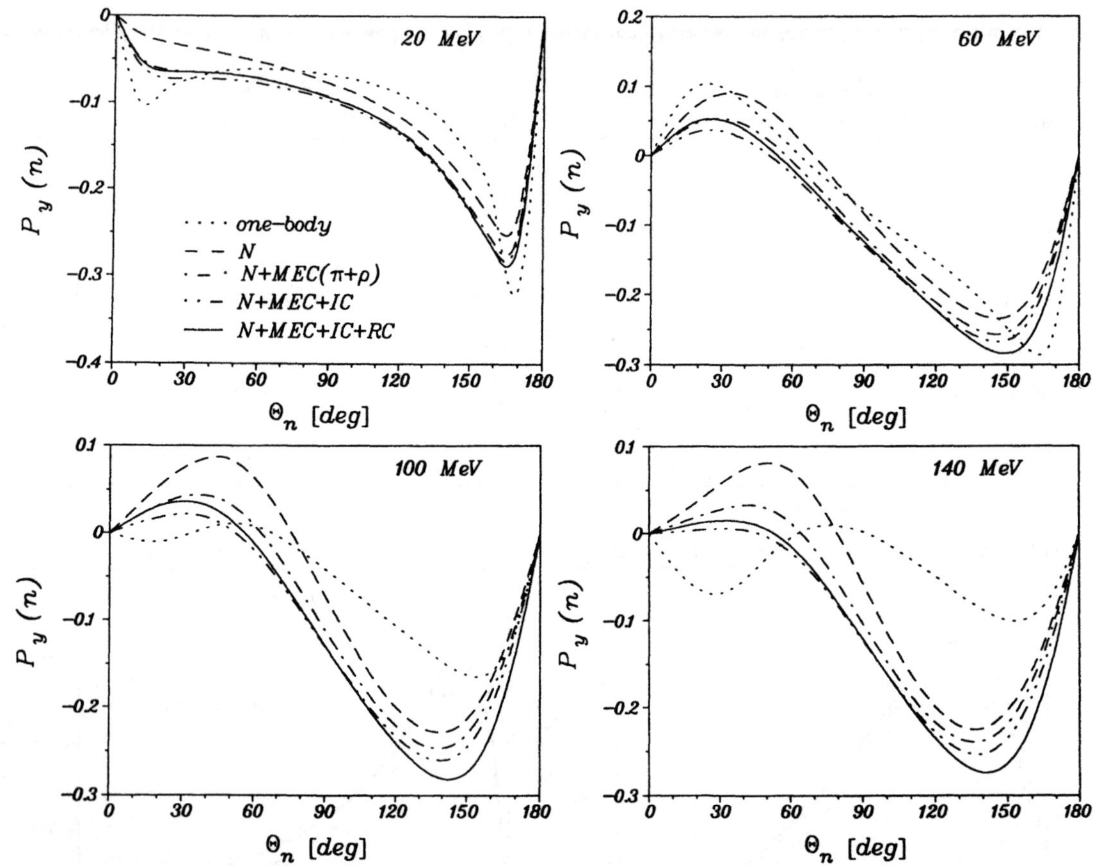

Fig. 7.4.5: Neutron polarization $P_y(n)$ at $E_\gamma = 20$, 60, 100, and 140 MeV with various contributions from MEC, IC, and RC calculated with the Bonn r-space potential.

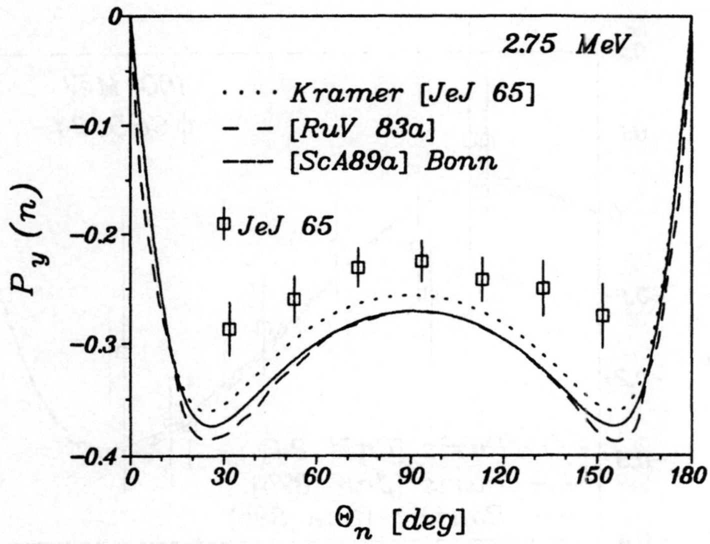

Fig. 7.4.6: Comparison of neutron polarization data on $P_y(n)$ with theoretical results at $E_\gamma = 2.75$ MeV.

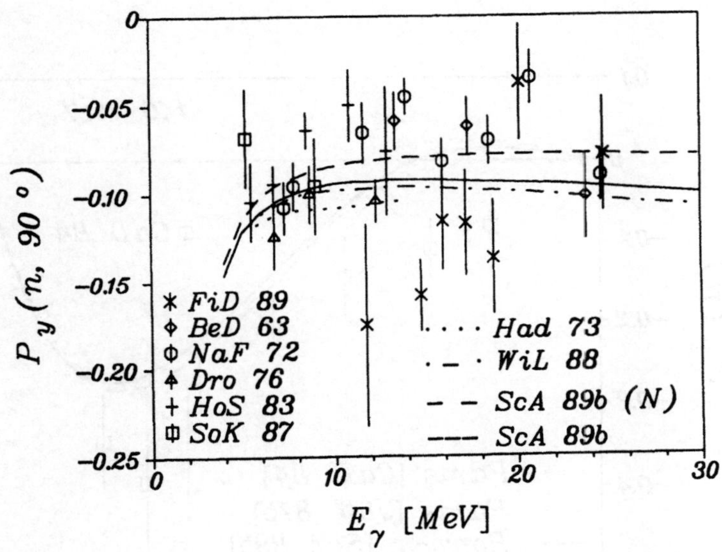

Fig. 7.4.7: Experimental data on neutron polarization $P_y(n)$ at 90° cm angle below photon lab energy $E_\gamma = 25$ MeV and theoretical predictions.

interaction is completely included in a coupled-channel approach. On the other hand, the various diagrammatic approaches of OgK 80, AnC 81 and Lag 84 encounter major problems in particular around 90°.

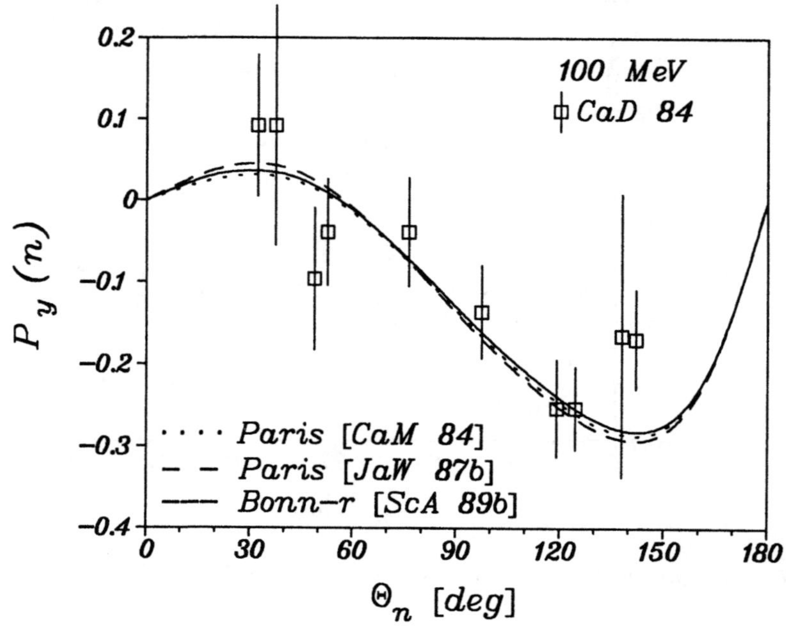

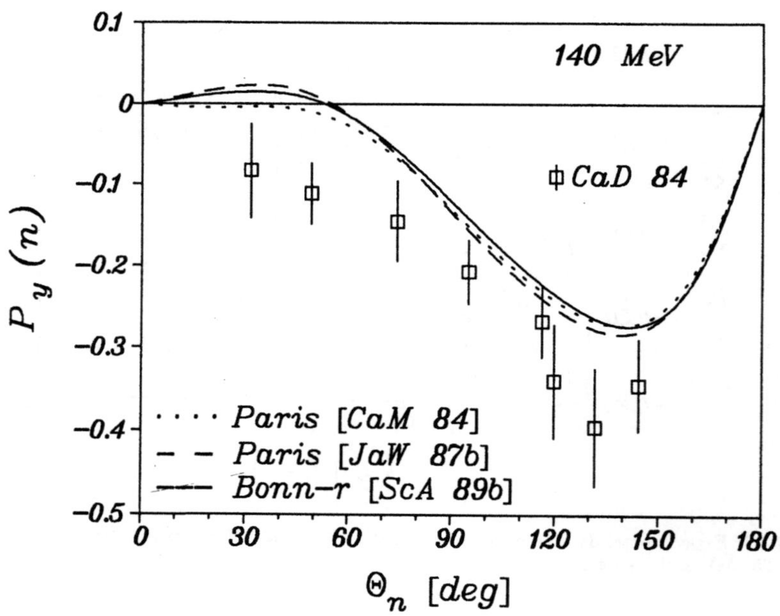

Fig. 7.4.8: Comparison of experimental data on neutron polarization $P_y(n)$ at $E_\gamma = 100$ and 140 MeV with theoretical results.

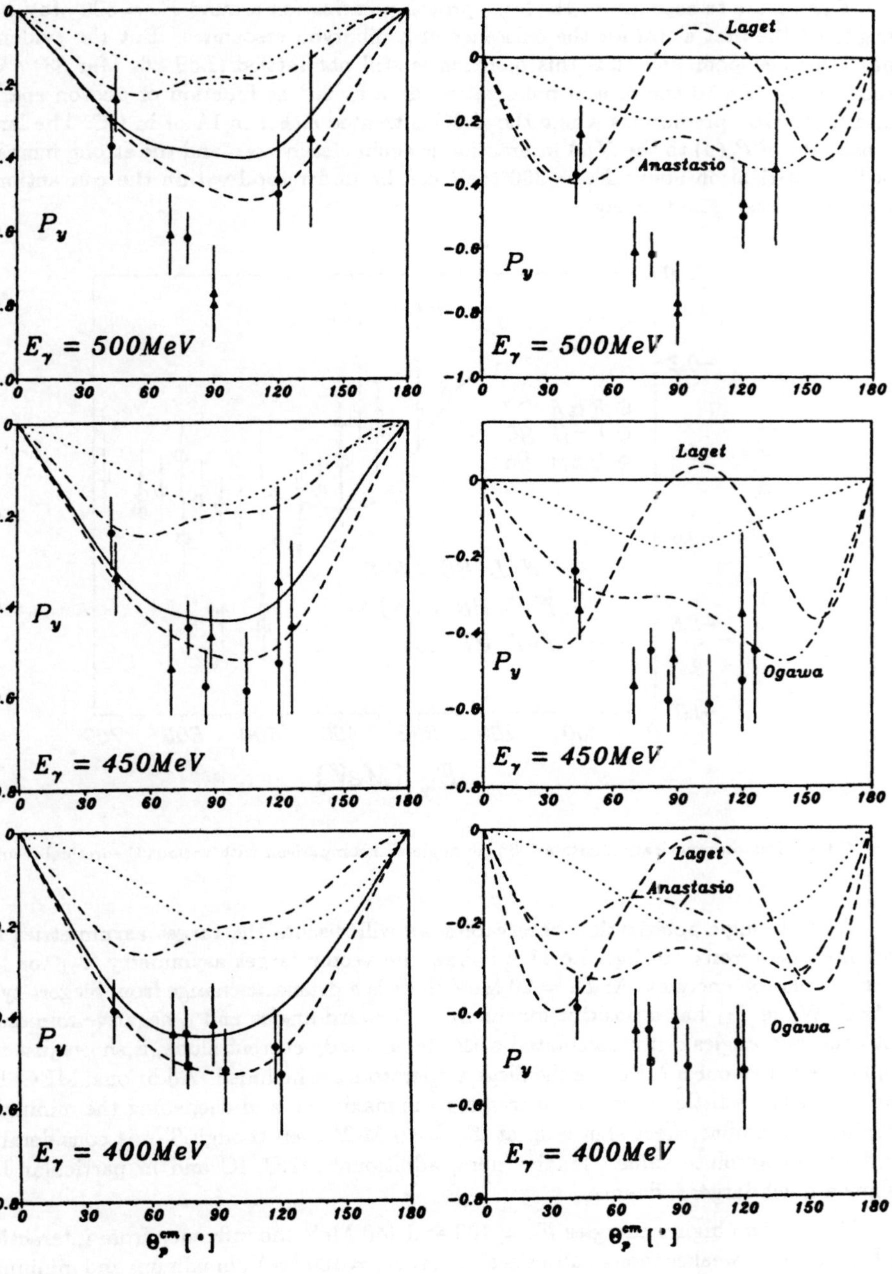

Fig. 7.4.9: Comparison of experimental data on proton polarization $P_y(p)$ at E_γ = 400, 450, and 500 MeV with various theoretical results (from LeA 87). Left: IC in IA (dotted) and IC in CC with different N-Δ interaction models (dashed, dashed-dotted, full). Right: IA from left (dotted) and results from OgK 80 (Ogawa), AnC 81 (Anastasio) and Lag 84 (Laget). Experimental data from LiL 68 (○), KoM 69 (□), KaA 77 (△), IkA 79 (△) and BrD 80 (○).

This failure to account for the large proton polarization around $E_\gamma = 500$ MeV had triggered the first claim for the existence of a dibaryon resonance. But the evidence has remained poor and thus this question is still not settled (LoS 86, Mar 88). We show in fig. 7.4.10 the proton polarization data at 90° as function of photon energy and theoretical predictions where the Δ-IC is treated either in IA or in CC. The large sensitivity of $P_y(p)$ to the N-Δ interaction is again clearly seen and the strong increase of the polarization above $E_\gamma = 300$ MeV can be understood within the conventional theory, at least qualitatively.

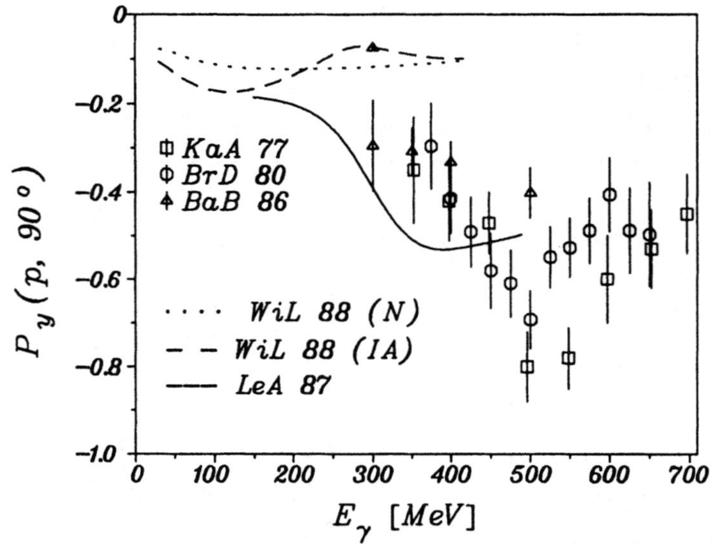

Fig. 7.4.10: Proton polarization data at 90° cm angle and comparison with various theoretical results.

As last single-polarization observables we will discuss the target asymmetries for oriented deuterons. In fig. 7.4.11 we show the vector target asymmetry T_{11} for the various photon energies. At $E_\gamma = 20$ MeV there is a dramatic change from Siegert-type MEC. While T_{11} has a positive maximum at forward angles and a negative minimum at backward angles if it is calculated with the one-body current alone, it shows just the opposite behaviour if MEC via the Siegert operators are included. Additional MEC, IC, and RC show a little effect only increasing the maximum and deepening the minimum slightly. A similar effect shows up at $E_\gamma = 60$ MeV even though T_{11} is considerably reduced in absolute value. Furthermore, additional MEC, IC and in particular RC show a much larger influence.

For the two higher energies $E_\gamma = 100$ and 140 MeV the influence from interaction effects is much weaker though still sizeable. At $E_\gamma = 100$ MeV, maximum and minimum of T_{11} are more and more reduced when one adds consecutively MEC, IC, and RC to the one-body current. In this case the strongest effect arises from relativistic corrections. On the other hand at $E_\gamma = 140$ MeV one finds a shift of the position of maximum and minimum and a noticeable increase of their absolute magnitude, which then, however, is again reduced by additional MEC, IC, and RC of almost equal size, so that at the

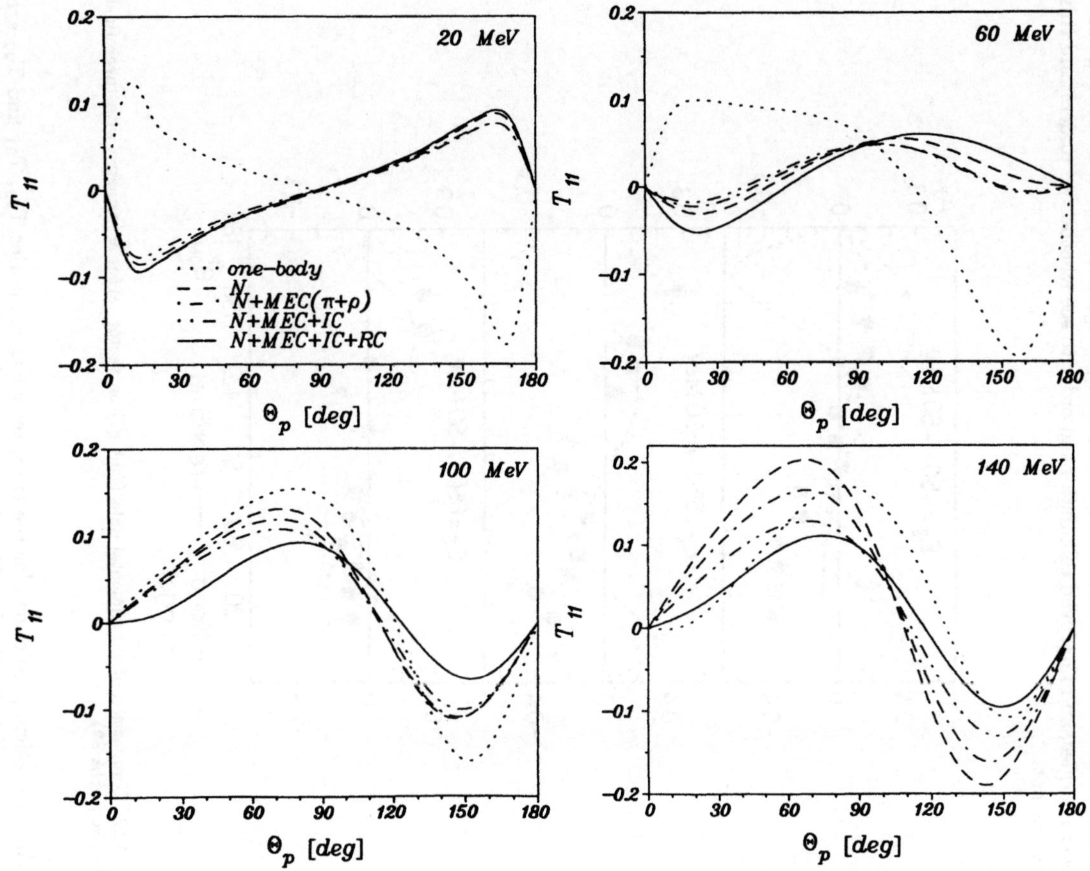

Fig. 7.4.11: Vector target asymmetry T_{11} at E_γ = 20, 60, 100, and 140 MeV with various contributions from MEC, IC, and RC calculated with the Bonn r-space potential.

end the maximum is much lower compared to the pure one-body result, whereas the minimum comes back close to the one without interaction effects.

Experimental data on the vector target asymmetry and several theoretical predictions are shown in fig. 7.4.12 which is taken from AlA 89. None of the theoretical calculations is able to describe quantitatively the data though the covariant approach of Kan 86 reproduces the qualitative behaviour fairly well. This is also true for the non-relativistic result of LeA 87 at the lowest energy. Again one needs more data, also at lower energies, and improved calculations in order to see whether we understand this process or not.

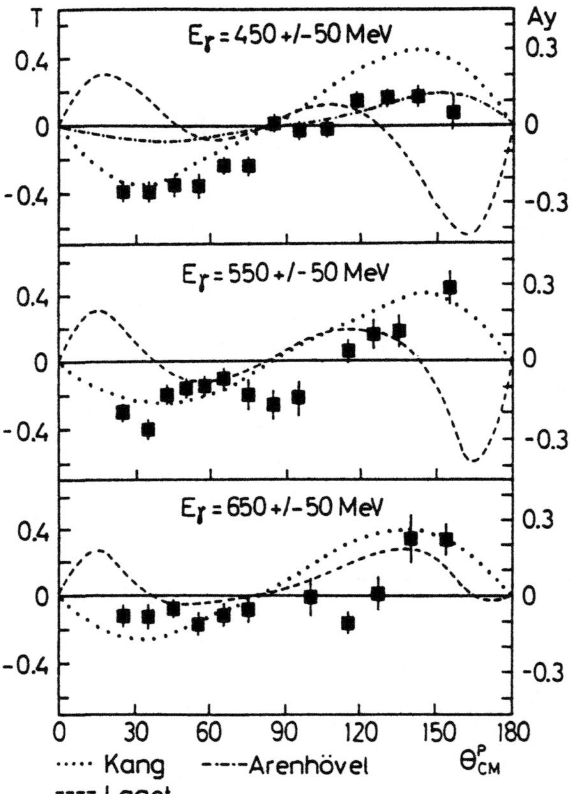

Fig. 7.4.12: Comparison of experimental data (AlA 89) on vector target asymmetry with theoretical results (from AlA 89).

The theoretical predictions for the three tensor asymmetries T_{20}, T_{21} and T_{22} are shown in figs. 7.4.13–15. For T_{20} the Siegert-type MEC show the largest effect at the lowest energy $E_\gamma = 20$ MeV in enhancing strongly T_{20} in forward and backward direction and reducing it around 90°, while with increasing photon energy the influence becomes less pronounced. Additional MEC and IC can be neglected for all four energies. But RC show at forward and backward angles a sizeable influence reducing T_{20} at 0° and

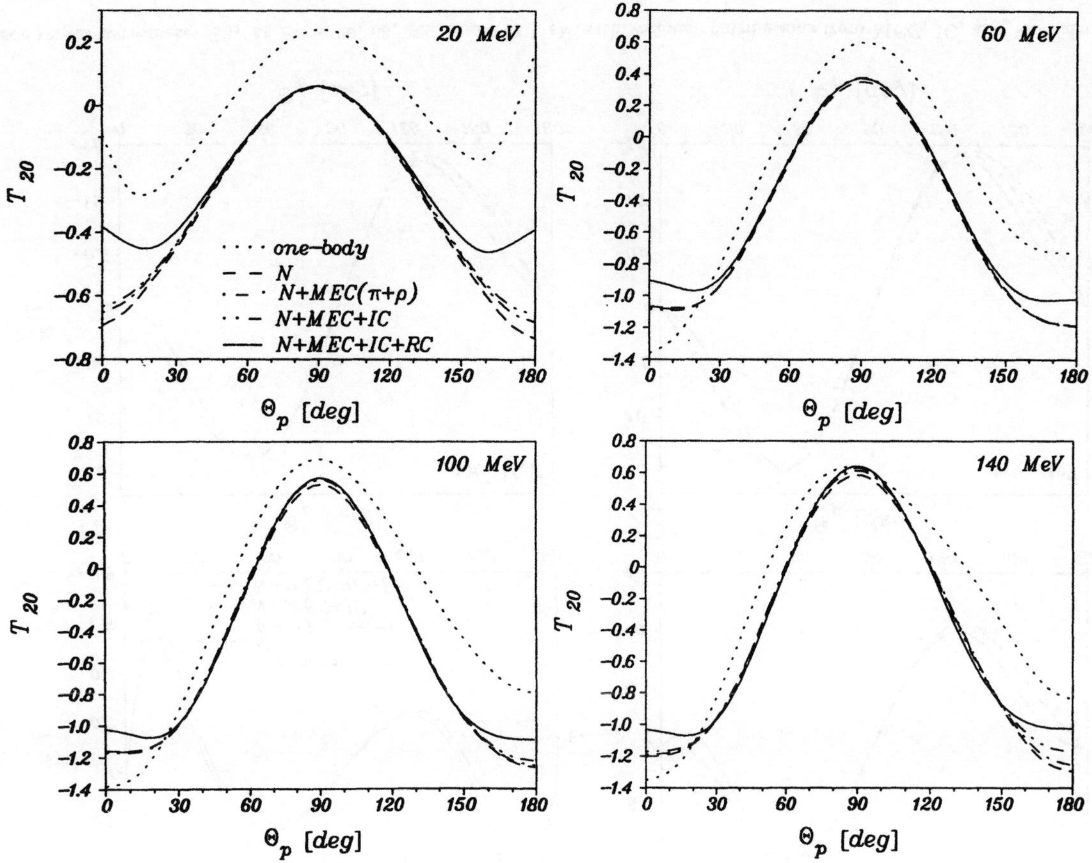

Fig. 7.4.13: Tensor target asymmetry T_{20} at E_γ = 20, 60, 100, and 140 MeV with various contributions from MEC, IC, and RC calculated with the Bonn r-space potential.

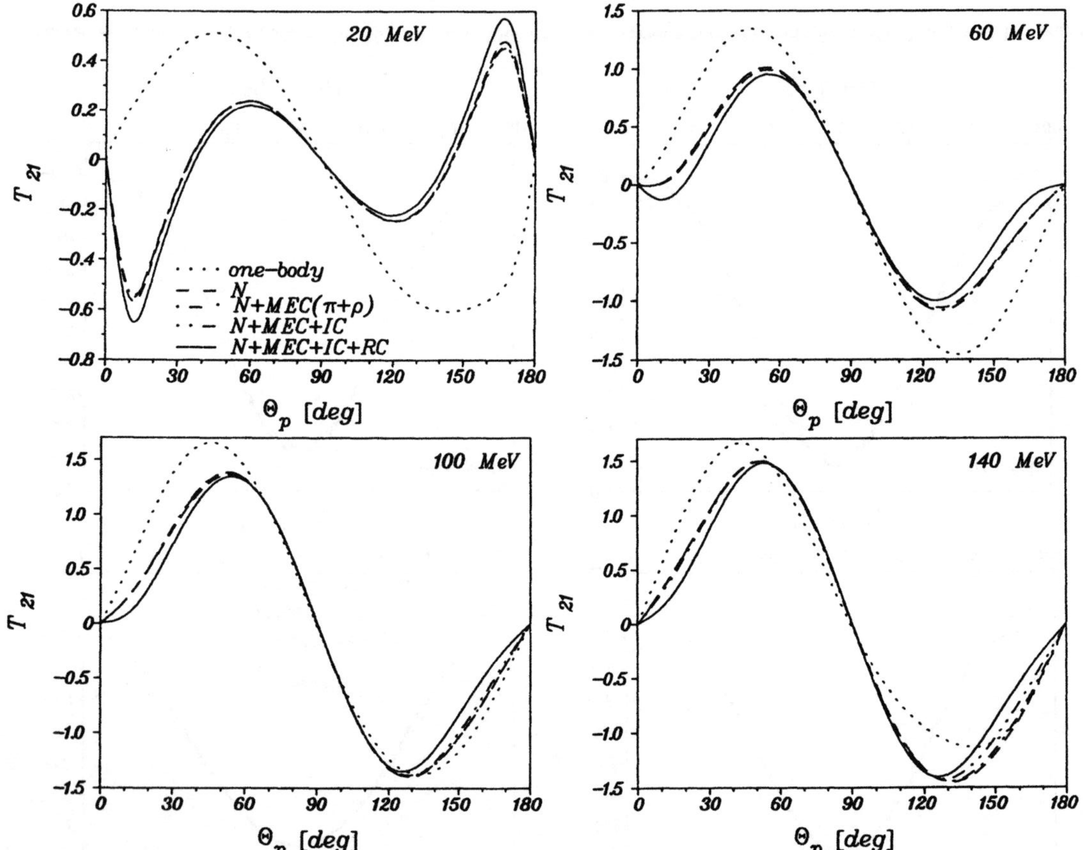

Fig. 7.4.14: Tensor target asymmetry T_{21} at $E_\gamma = 20$, 60, 100, and 140 MeV with various contributions from MEC, IC, and RC calculated with the Bonn r-space potential.

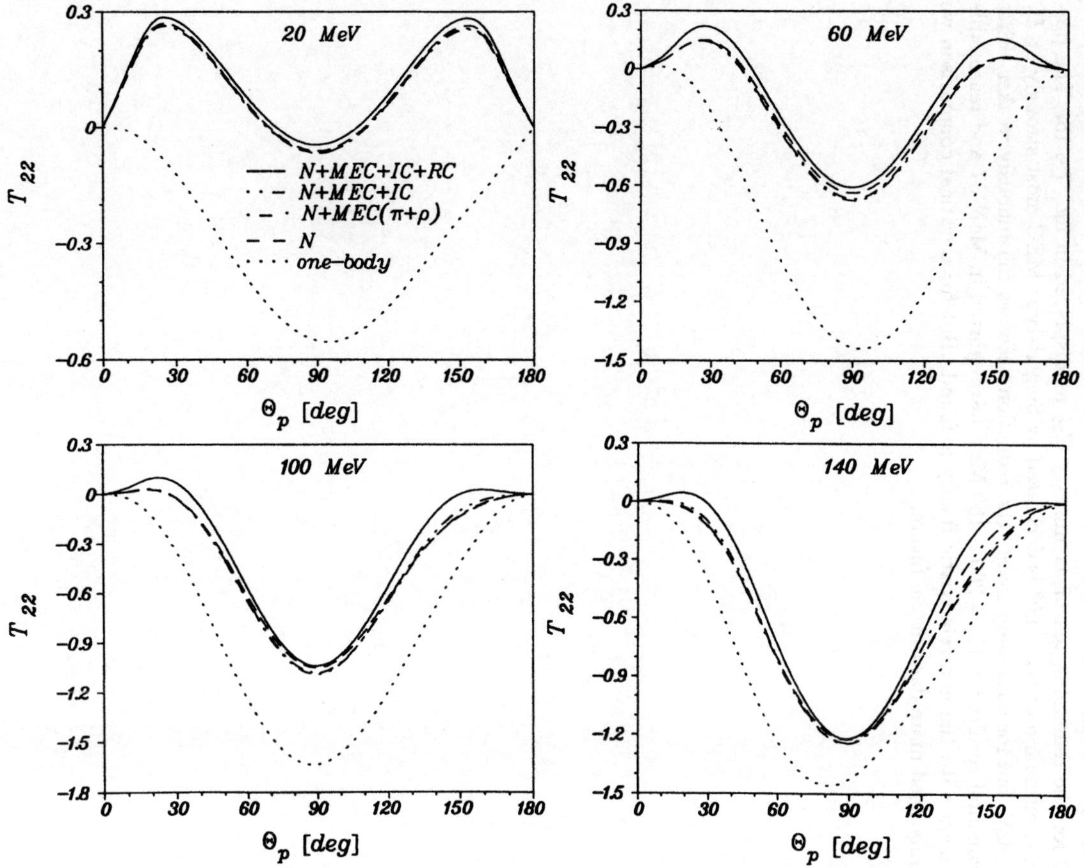

Fig. 7.4.15: Tensor target asymmetry T_{22} at E_γ = 20, 60, 100, and 140 MeV with various contributions from MEC, IC, and RC calculated with the Bonn r-space potential.

180° by almost a factor of 2 at $E_\gamma = 20$ MeV. This behaviour of relativistic corrections is similar to what has been found for the differential cross section as discussed in section 7.3.

The tensor asymmetry T_{21} in fig. 7.4.14 is mainly affected by Siegert-type MEC with most drastic changes at $E_\gamma = 20$ MeV. For the other higher energies the effect is much weaker. For all energies additional MEC and IC show again almost no influence. Only RC give a small effect at forward and backward angles. A similar behaviour is found for the remaining tensor asymmetry T_{22} as displayed in fig. 7.4.15. The pronounced minimum at $\Theta_p = 90°$ is decreased by Siegert-type MEC most strongly at $E_\gamma = 20$ MeV, but for increasing energy the reduction becomes continuously weaker. First experimental results for $E_\gamma = 200 - 160$ MeV are reported in MoN 87 and reasonable agreement with the non-relativistic theory is found. But for a critical comparison we certainly need more data in the future.

8 Conclusion and outlook

The present status of experimental and theoretical investigations of deuteron photodisintegration has been thoroughly reviewed. The various experimental techniques for measuring total and differential cross sections and as well as polarization observables have been described and the existing experimental data have been critically evaluated and selected in order to provide a sound basis for a comparison with theoretical results.

The lack of an accurate knowledge of the incoming photon flux was the main reason of the large discrepancies, up to 50 percent, among the experimental results on the absolute cross section, obtained with bremsstrahlung photon beams. During the last decade several methods have been applied in order to have monochromatic photons or to measure accurately the photon flux. Just by selecting these experiments it was possible to reduce drastically the discrepancy and to provide an experimental data set on differential and total cross sections, which lies in a band of relative width of a few percent at low energy and of about 20% in the Δ region.

In order to obtain more accurate results a further effort must be done towards a better evaluation of the detector efficiency, the effective solid angle and the effective thickness of the deuteron target, a better control of the stability of the whole apparatus and a more accurate measurement of the background contribution.

Further information has been obtained by relative quantities like the relative angular distribution, the ratios between the differential cross sections measured at two angles, the photon asymmetry and the analyzing power. These quantities are independent of the knowledge of the photon flux and we have seen that the agreement between the experimental data concerning these quantities is in general good and the uncertainty is restricted to a maximum value of 10 percent. It must be noted that they are in any case affected by the other systematic errors still present also in these measurements.

Finally, information has been extracted by the ensemble of the data on differential and total cross sections by averaging the weighted data taking into account both statistical and systematic errors when known. In this way some characteristic gross features of the cross section are now experimentally well known, like the position and the width of the Δ resonance, and the behaviour of the angular distribution and of the photon asymmetry at $\Theta_\gamma = 90°$ with the energy, at least up to the Δ region.

The various concepts underlying the different theoretical approaches for the calculation of the observables have been outlined and discussed in detail. Specific emphasis has been laid on the role of subnuclear degrees of freedom in terms of mesons and isobars and on the influence of relativistic effects. The framework which has been used for the theoretical discussion is based on a model space with explicit meson, nucleon, and isobar degrees of freedom.

In the non-relativistic approach the mesons are considered in terms of effective operators like the N-N interaction and meson-exchange currents. It is very important that these operators are constructed in a consistent manner in order to fulfil current conservation as a gauge constraint. Modern models for realistic N-N interactions give a reliable overall description of the hadronic structure of the bound and scattering states of the p-n system in the long- and intermediate-range region, whereas in the short-range part a large uncertainty remains.

The dominant role of the lowest-order isovector π-MEC is well established. It is related to the longest-range static one-pion-exchange part of the N-N interaction. In general exchange currents affect most strongly $E1$ transitions. However, the major part is already included if the Siegert operator is used. At low energies isobar configurations contribute little but become more important with increasing energy and dominate in the Δ region. Here a careful treatment of the N-Δ interaction is necessary.

Relativistic effects show a clear signature already at low energies for forward and backward nucleon emission with respect to the incoming-photon direction. In general they are treated as lowest-order corrections to wave functions and operators. But a consistent theory including a readjustment of the N-N interaction is still missing.

In comparing the various theoretical concepts, as outlined in section 4, it is fair to say that for photon energies up to about the pion threshold the non-relativistic approach using realistic potentials as input and including MEC, IC, and RC of lowest order gives the most reliable results. Small inconsistencies in the MEC operators do not matter if the Siegert operators are used. Up to about $E_\gamma = 100$ MeV the dependence on the N-N interaction model of the differential cross section and the most important polarization observables is rather weak, in general up to a few percent which is still less than present experimental uncertainties. Thus there remains only very little room for theoretical modification within this approach.

In this energy region the diagrammatic approaches have not reached such a high degree of accuracy because the final-state interaction is treated only approximately and the initial deuteron wave function is taken from a non-relativistic model, which introduces an additional inconsistency. It is further aggravated by lack of consistency in the meson-exchange current operators.

Around and above pion threshold major theoretical uncertainties arise from the different treatments of Δ degrees of freedom with respect to off-shell effects, to the NN-$N\Delta$ transition interaction and the $N\Delta$ transition current and whether the isobar configurations are calculated in impulse approximation or in a coupled-channel approach. Also the coupling to the $NN\pi$ channel needs in general a better treatment. Only in the unitary three-body model this has been done. In addition, relativistic effects are expected to become even more important at these energies and thus require a careful and consistent consideration. Therefore, one finds a much larger variation of theoretical results in this energy region and at present it is impossible to prefer one theoretical approach against another. This conclusion is also valid for the attempt to improve the theoretical description of the differential cross section above the Δ-resonance region by including a hypothetical dibaryon resonance (Mak 84). The comparison of the few available data with calculations is very ambiguous. Future theoretical efforts should concentrate on an improved description in this energy region. Also the consequences of QCD concerning the possible role of quark-gluon dynamics will have an important impact on future studies.

Within present experimental accuracy one finds good agreement of theoretical total cross sections with the data at low and intermediate energies up to about $E_\gamma = 100$ MeV. At higher energies including the Δ region the agreement can still be considered satisfactory though larger experimental errors and greater theoretical uncertainties and model dependences prevent to draw definite conclusions whether or not the present theoretical framework suffices to describe this process.

Angular distributions allow in principle a more sensitive test of the theory, provided one has good experimental data. But even if, as has been discussed in detail, a considerable improvement by careful selection and averaging of the data is achieved, the residual uncertainties still remain large. Nevertheless the comparison between experimental and theoretical values, in particular for the ratio $\frac{d\sigma(0°)}{d\Omega} / \frac{d\sigma(90°)}{d\Omega}$ and the Legendre polynomial coefficient A_2, shows clearly that at present the theory is not able to reproduce the behaviour of the differential cross section in the whole Δ region. Without doubt, we need more data in the higher energy range with an accuracy of a few percent in order to find out how much is the real discrepancy between theory and experiment and to stimulate further calculations.

The study of polarization observables will constitute another promising future field of research because these observables provide us with additional sensitive tools for testing models of subnuclear degrees of freedom. The present situation concerning the comparison of experimental data on polarization observables with theoretical predictions is inconclusive. Of the few polarization observables for which data exists, the photon asymmetry for linearly polarized photons is in good agreement with theory at least for low and intermediate energies. Larger discrepancies seem to exist at high energies. In particular the experimental data show clearly a sign change at about 550 MeV while in the theoretical evaluation the sign change is around 400 MeV (see fig. 7.4.3). With respect to vector and tensor target asymmetries, mainly data for T_{11} exist but at high energy only, where the theoretical uncertainties are large.

The situation for the nucleon polarization is even worse. Already at low energies the few data points do not allow a definite conclusion as to agreement or disagreement with the conventional theories. And in the high-energy region, where the possible evidence of dibaryon resonances has been discussed, the theoretical predictions depend sensitively on dynamical details like, for example, the N-Δ interaction model. Up till now deuteron photodisintegration does not give any convincing evidence for the existence of dibaryon resonances. The differences between experimental data and theoretical calculations are of the same order of magnitude as the variation between different calculations, independent of the fact whether dibaryon resonances are included or not. In principle, this process could serve as an important test for the existence of such resonances provided that precise experimental data and reliable theories are available.

Thus the exploitation of polarization observables in order to deepen our understanding of this important process has to be considered of being in the initial stages only and, therefore, a large fraction of future research should be devoted to a careful and more systematic study of a greater number of polarization observables, in particular double- and triple-polarization observables.

After fifty-five years the photodisintegration of deuteron is still open to experimental and theoretical investigations. Despite the large production of experimental and theoretical papers a better description of the two-body photodisintegration of the deuteron above the pion production threshold has still to be achieved. Even though the reliability of recent experimental data, using monochromatic or "quasi-monochromatic" photons has been improved, the accuracy is still not satisfactory and more systematic measurements are needed with a few percent accuracy.

Anyway, at present there are no calculations available which describe the whole set of information which one can deduce from the experiments. The dependence of some observables on the potential model is rather large and at present there is not one potential available which gives the best description of all the measured quantities.

Moreover, one important question has not been answered at the moment: Will the conventional theory with nucleon meson and isobar d.o.f. alone be able to describe the experimental results or are some basic ingredients of fundamental importance (like quark-gluon d.o.f.), which are still missing in the current theory? This would not be surprising, since at energies above the Δ region the wave length of the photon becomes comparable to the expected distance between quarks in the nucleus.

Appendix A: Nucleon polarization observables

For the polarization components of one outgoing nucleon ($j = 1, 2$) one has

$$
\begin{aligned}
P_{x/z}(j)\frac{d\sigma}{d\Omega} &= \frac{d\sigma_0}{d\Omega}\sum_{I=0}^{2} P_I^d\Big[\sum_{M=0}^{I}\big(P_{x/z}^{0,IM}(j,\Theta)\sin(M(\Phi_d-\Phi)+\delta_{I1}\pi/2)\\
&\quad + P_c^\gamma P_{x/z}^{c,IM}(j,\Theta)\cos(M(\Phi_d-\Phi)-\delta_{I1}\pi/2)\big)d_{M0}^I(\Theta_d)\\
&\quad + P_l^\gamma\sum_{M=-I}^{I} P_{x/z}^{l,IM}(j,\Theta)\sin(\psi_M+\delta_{I1}\pi/2)d_{M0}^I(\Theta_d)\Big] ,
\end{aligned} \tag{A.1}
$$

$$
\begin{aligned}
P_y(j)\frac{d\sigma}{d\Omega} &= \frac{d\sigma_0}{d\Omega}\sum_{I=0}^{2} P_I^d\Big[\sum_{M=0}^{I}\big(P_y^{0,IM}(j,\Theta)\cos(M(\Phi_d-\Phi)-\delta_{I1}\pi/2)\\
&\quad + P_c^\gamma P_y^{c,IM}(j,\Theta)\sin(M(\Phi_d-\Phi)+\delta_{I1}\pi/2)\big)d_{M0}^I(\Theta_d)\\
&\quad + P_l^\gamma\sum_{M=-I}^{I} P_y^{l,IM}(j,\Theta)\cos(\psi_M-\delta_{I1}\pi/2)d_{M0}^I(\Theta_d)\Big] ,
\end{aligned} \tag{A.2}
$$

where the observables are given by

$$
P_{x/y}^{0,IM}(j,\Theta)\frac{d\sigma_0}{d\Omega} = \begin{cases} -\mathrm{Im}\ Y_{IM}^{\mp}(j)\ , & I=0,2\ ,\\ \mp\mathrm{Re}\ Y_{1M}^{\pm}(j)\ , & I=1\ ,\end{cases} \tag{A.3}
$$

$$
P_{x/y}^{c,IM}(j,\Theta)\frac{d\sigma_0}{d\Omega} = \begin{cases} \mp\mathrm{Re}\ Y_{IM}^{\pm}(j)\ , & I=0,2\ ,\\ -\mathrm{Im}\ Y_{1M}^{\mp}(j)\ , & I=1\ ,\end{cases} \tag{A.4}
$$

$$
P_{x/y}^{l,IM}(j,\Theta)\frac{d\sigma_0}{d\Omega} = -\begin{cases} iZ_{IM}^{\mp}(j)\ , & I=0,2\ ,\\ \mp Z_{1M}^{\mp}(j)\ , & I=1\ ,\end{cases} \tag{A.5}
$$

$$
P_{z}^{0,IM}(j,\Theta)\frac{d\sigma_0}{d\Omega} = \frac{2\sqrt{2}}{1+\delta_{M0}}\begin{cases} \mathrm{Im}\ Y_{IM}^{0}(j)\ , & I=0,2\ ,\\ \mathrm{Re}\ Y_{1M}^{0}(j)\ , & I=1\ ,\end{cases} \tag{A.6}
$$

$$
P_{z}^{c,IM}(j,\Theta)\frac{d\sigma_0}{d\Omega} = \frac{2\sqrt{2}}{1+\delta_{M0}}\begin{cases} \mathrm{Re}\ Y_{IM}^{0}(j)\ , & I=0,2\ ,\\ \mathrm{Im}\ Y_{1M}^{0}(j)\ , & I=1\ ,\end{cases} \tag{A.7}
$$

$$
P_{z}^{l,IM}(j,\Theta)\frac{d\sigma_0}{d\Omega} = \sqrt{2}\begin{cases} iZ_{IM}^{0}(j)\ , & I=0,2\ ,\\ -Z_{1M}^{0}(j)\ , & I=1\ ,\end{cases} \tag{A.8}
$$

with

$$
Y_{IM}^{\pm}(j) = \frac{2}{1+\delta_{M0}}\left(Y_{IM}^{1}(j)\pm(-)^M Y_{I-M}^{1}(j)\right) , \tag{A.9}
$$

$$
Z_{IM}^{\pm}(j) = Z_{IM}^{1}(j)\pm Z_{IM}^{-1}(j) \tag{A.10}
$$

and ($\rho = 0$ and ±1)

$$
Y/Z_{IM}^{\rho}(1) = \sum_{s's}(-)^s\hat{s}'\hat{s}\begin{Bmatrix} s' & s & 1\\ \frac{1}{2} & \frac{1}{2} & \frac{1}{2}\end{Bmatrix} v/w_{IM}^{s's1\rho} , \tag{A.11}
$$

for particle "1", and $Y^{\rho}_{IM}(2)$ and $Z^{\rho}_{IM}(2)$ are analogously defined with an additional phase factor $(-)^{s'+s}$. Note that for particle "2" the same coordinate system as for particle "1" has been used.

The vanishing of the polarizations P_x and P_z for unpolarized photons and deuterons is a consequence of parity conservation because they are expectation values of pseudoscalars $\boldsymbol{\sigma}\cdot((\hat{\boldsymbol{\omega}}\times\hat{\boldsymbol{k}})\times\hat{\boldsymbol{\omega}})$ and $\boldsymbol{\sigma}\cdot\hat{\boldsymbol{\omega}}$, respectively, while P_y corresponds to a scalar $\boldsymbol{\sigma}\cdot(\hat{\boldsymbol{\omega}}\times\hat{\boldsymbol{k}})$.

For the polarization of both outgoing nucleons one finds

$$\begin{aligned} P_{x_ix_j}(1,2)\frac{d\sigma}{d\Omega} &= \frac{d\sigma_0}{d\Omega}\sum_{I=0}^{2} P_I^d\,[\sum_{M=0}^{I}(P^{0,IM}_{x_ix_j}\cos(M(\Phi_d-\Phi)-\delta_{I1}\pi/2) \\ &\quad + P_c^\gamma P^{c,IM}_{x_ix_j}\sin(M(\Phi_d-\Phi)+\delta_{I1}\pi/2))d^I_{M0}(\Theta_d) \\ &\quad + P_l^\gamma\sum_{M=-I}^{I} P^{l,IM}_{x_ix_j}\cos(\psi_M-\delta_{I1}\pi/2)d^I_{M0}(\Theta_d)]\,, \end{aligned} \tag{A.12}$$

for the combinations

$$x_ix_j \;=\; xx,\; yy,\; zz,\; xz,\; zx, \tag{A.13}$$

where

$$P^{0,IM}_{x_ix_j}(1,2)\frac{d\sigma_0}{d\Omega} = \begin{cases}\mathrm{Re}\,\bar{Y}_{IM}(x_ix_j)\,, & I=0,2\,,\\ \mathrm{Im}\,\bar{Y}_{1M}(x_ix_j)\,, & I=1\,,\end{cases} \tag{A.14}$$

$$P^{c,IM}_{x_ix_j}(1,2)\frac{d\sigma_0}{d\Omega} = \begin{cases}\mathrm{Im}\,\bar{Y}_{IM}(x_ix_j)\,, & I=0,2\,,\\ \mathrm{Re}\,\bar{Y}_{1M}(x_ix_j)\,, & I=1\,,\end{cases} \tag{A.15}$$

$$P^{l,IM}_{x_ix_j}(1,2)\frac{d\sigma_0}{d\Omega} = \begin{cases}-\bar{Z}_{IM}(x_ix_j)\,, & I=0,2\,,\\ i\bar{Z}_{1M}(x_ix_j)\,, & I=1\,,\end{cases} \tag{A.16}$$

with

$$\bar{Y}_{IM}(xx/yy) = -\frac{4}{1+\delta_{M0}}\left(\bar{Y}^{00}_{IM}+\frac{1}{\sqrt{2}}\bar{Y}^{20}_{IM}\mp\frac{\sqrt{3}}{2}(\bar{Y}^{22}_{IM}+\bar{Y}^{2-2}_{IM})\right)\,, \tag{A.17}$$

$$\bar{Y}_{IM}(zz) = -\frac{4}{1+\delta_{M0}}(\bar{Y}^{00}_{IM}-\sqrt{2}\bar{Y}^{20}_{IM})\,, \tag{A.18}$$

$$\bar{Y}_{IM}(xz/zx) = -\frac{2\sqrt{3}}{1+\delta_{M0}}(\pm(\bar{Y}^{11}_{IM}+\bar{Y}^{1-1}_{IM})+\bar{Y}^{21}_{IM}-\bar{Y}^{2-1}_{IM})\,, \tag{A.19}$$

$$\bar{Z}_{IM}(xx/yy) = -2\left(\bar{Z}^{00}_{IM}+\frac{1}{\sqrt{2}}\bar{Z}^{20}_{IM}\mp\frac{\sqrt{3}}{2}(\bar{Z}^{22}_{IM}+\bar{Z}^{2-2}_{IM})\right)\,, \tag{A.20}$$

$$\bar{Z}_{IM}(zz) = -2(\bar{Z}^{00}_{IM}-\sqrt{2}\bar{Z}^{20}_{IM})\,, \tag{A.21}$$

$$\bar{Z}_{IM}(xz/zx) = -\sqrt{3}(\pm(\bar{Z}^{11}_{IM}+\bar{Z}^{1-1}_{IM})+\bar{Z}^{21}_{IM}-\bar{Z}^{2-1}_{IM})\,, \tag{A.22}$$

and

$$\bar{Y}/\bar{Z}^{S\sigma}_{IM} \;=\; \hat{S}\sum_{s's}\hat{s}'\hat{s}\begin{Bmatrix}\frac{1}{2} & \frac{1}{2} & 1\\ \frac{1}{2} & \frac{1}{2} & 1\\ s' & s & S\end{Bmatrix} v/w^{s'sS\sigma}_{IM}\,. \tag{A.23}$$

For the remaining combinations

$$x_i x_j = xy,\ yx,\ zy,\ yz \tag{A.24}$$

one has

$$\begin{aligned} P_{x_i x_j}(1,2)\frac{d\sigma}{d\Omega} = {} & \frac{d\sigma_0}{d\Omega}\sum_{I=0}^{2} P_I^d \, [\sum_{M=0}^{I} (P_{x_i x_j}^{0,IM} \sin(M(\Phi_d - \Phi) + \delta_{I1}\pi/2) \\ & + P_c^\gamma P_{x_i x_j}^{c,IM} \cos(M(\Phi_d - \Phi) - \delta_{I1}\pi/2)) d_{M0}^I(\Theta_d) \\ & + P_l^\gamma \sum_{M=-I}^{I} P_{x_i x_j}^{l,IM} \sin(\psi_M + \delta_{I1}\pi/2) d_{M0}^I(\Theta_d)] \,, \end{aligned} \tag{A.25}$$

with

$$P_{x_i x_j}^{0,IM}\frac{d\sigma_0}{d\Omega} = \begin{cases} \text{Im } \bar{Y}_{1M}(x_i x_j)\,, & I = 1\,, \\ -\text{Re } \bar{Y}_{2M}(x_i x_j)\,, & I = 2\,, \end{cases} \tag{A.26}$$

$$P_{x_i x_j}^{c,IM}\frac{d\sigma_0}{d\Omega} = \begin{cases} \text{Im } \bar{Y}_{IM}(x_i x_j)\,, & I = 0,2\,, \\ -\text{Re } \bar{Y}_{1M}(x_i x_j)\,, & I = 1\,, \end{cases} \tag{A.27}$$

$$P_{x_i x_j}^{l,IM}\frac{d\sigma_0}{d\Omega} = \begin{cases} \bar{Z}_{IM}(x_i x_j)\,, & I = 0,2\,, \\ i\bar{Z}_{1M}(x_i x_j)\,, & I = 1\,, \end{cases} \tag{A.28}$$

and

$$\bar{Y}_{IM}(xy/yx) = -\frac{2\sqrt{3}}{1+\delta_{M0}}(\pm\sqrt{2}\bar{Y}_{IM}^{10} + \bar{Y}_{IM}^{22} - \bar{Y}_{IM}^{2-2})\,, \tag{A.29}$$

$$\bar{Y}_{IM}(yz/zy) = -\frac{2\sqrt{3}}{1+\delta_{M0}}(\pm(\bar{Y}_{IM}^{11} - \bar{Y}_{IM}^{1-1}) + \bar{Y}_{IM}^{21} + \bar{Y}_{IM}^{2-1})\,, \tag{A.30}$$

$$\bar{Z}_{IM}(xy/yx) = \sqrt{3}(\pm\sqrt{2}\bar{Z}_{IM}^{10} + \bar{Z}_{IM}^{22} - \bar{Z}_{IM}^{2-2})\,, \tag{A.31}$$

$$\bar{Z}_{IM}(yz/zy) = \sqrt{3}(\pm(\bar{Z}_{IM}^{11} - \bar{Z}_{IM}^{1-1}) - \bar{Z}_{IM}^{21} - \bar{Z}_{IM}^{2-1})\,. \tag{A.32}$$

The vanishing of $P_{x_i x_j}$ for unpolarized photons and deuterons for the combinations (A.24) is again a consequence of parity conservation as mentioned for $P_{x/z}$ before.

Appendix B: Relations between the coefficients of the de Swart-Partovi and the Legendre polynomial expansion of the differential cross section

For the differential cross section one finds the relations between the coefficients by equating the two expansions

$$\sum_K V_{00}^K P_k(\cos\Theta) = \sum_{n=0} (a_n + b_n \cos\Theta) \sin^{2n}\Theta . \qquad (B.1)$$

This gives the general relations (CaM 82a)

$$a_n = \frac{(-)^n}{2^n (n!)^2} \sum_{K=n}^{\infty} \frac{K!}{(K-n)!} \frac{(2(K+n)-1)!!}{(2K-1)!!} V_{00}^{2K} \qquad (B.2)$$

$$b_n = \frac{(-)^n}{2^n (n!)^2} \sum_{K=n}^{\infty} \frac{K!}{(K-n)!} \frac{(2(K+n)+1)!!}{(2K+1)!!} V_{00}^{2K+1} . \qquad (B.3)$$

Explicitly one finds

$$a_0 = \sum_{K=0}^{\infty} V_{00}^{2K} , \qquad (B.4)$$

$$\begin{aligned} a_1 &= -\frac{1}{2} \sum_{K=1}^{\infty} K(2K+1) V_{00}^{2K} \\ &= -\frac{1}{2}(3V_{00}^2 + 10V_{00}^4 + 21V_{00}^6 + 36V_{00}^8 + 55V_{00}^{10} + \ldots) , \end{aligned} \qquad (B.5)$$

$$\begin{aligned} a_2 &= \frac{1}{16} \sum_{K=2}^{\infty} K(K-1)(2K+1)(2K+3) V_{00}^{2K} \\ &= \frac{1}{8}(35V_{00}^4 + 189V_{00}^6 + 594V_{00}^8 + 1430V_{00}^{10} + \ldots) , \end{aligned} \qquad (B.6)$$

$$b_0 = \sum_{K=0}^{\infty} V_{00}^{2K+1} , \qquad (B.7)$$

$$\begin{aligned} b_1 &= -\frac{1}{2} \sum_{K=1}^{\infty} K(2K+3) V_{00}^{2K+1} \\ &= -\frac{1}{2}(5V_{00}^3 + 14V_{00}^5 + 27V_{00}^7 + 44V_{00}^9 + 65V_{00}^{11} + \ldots) , \end{aligned} \qquad (B.8)$$

$$\begin{aligned} b_2 &= \frac{1}{16} \sum_{K=2}^{\infty} K(K-1)(2K+3)(2K+5) V_{00}^{2K+1} \\ &= \frac{1}{8}(63V_{00}^5 + 297V_{00}^7 + 858V_{00}^9 + 1950V_{00}^{11} + \ldots) . \end{aligned} \qquad (B.9)$$

In the notation of de Swart (DSw 59) one has

$$\begin{aligned} a_0 &= a \qquad & a_1 &= b \\ b_0 &= a\beta_1 \qquad & b_1 &= b\beta_2 \end{aligned} \qquad (B.10)$$

and in the notation of Partovi (Par 64)

$$\begin{array}{lll} a_0 = a & a_1 = b & a_2 = e \\ b_0 = c & b_1 = d\,. & \end{array} \qquad (B.11)$$

Appendix C: Expansion of nucleon polarization observables in terms of Legendre functions

(i) One-nucleon polarization: Since all observables are given by the real or imaginary part of $Y_{IM}^{0,\pm}(j)$ and $Z_{IM}^{0,\pm}(j)$ it is sufficient to write down their expansions

$$Y_{IM}^{0}(j,\Theta) = \sum_K y_{IM0}^K(j) d_{M0}^K(\Theta) , \tag{C.1}$$

$$Y_{IM}^{\pm}(j,\Theta) = \frac{2}{1+\delta_{M0}} \sum_K (y_{IM1}^K(j) d_{M,-1}^K(\Theta) \mp y_{I-M1}^K d_{M1}^K(\Theta)) , \tag{C.2}$$

$$Z_{IM}^{0}(j,\Theta) = \sum_K z_{IM0}^K(j) d_{M-2,0}^K(\Theta) , \tag{C.3}$$

$$Z_{IM}^{\pm}(j,\Theta) = \sum_K (z_{IM1}^K(j) d_{M-2,-1}^K(\Theta) \pm z_{IM-1}^K(j) d_{M-2,1}^K(\Theta)) , \tag{C.4}$$

where the coefficients are defined by

$$y/z_{IM\rho}^K(1) = (-)^{1+M} \sum_{s's} (-)^{s'} \hat{s}' \hat{s} \begin{Bmatrix} s' & s & 1 \\ \frac{1}{2} & \frac{1}{2} & \frac{1}{2} \end{Bmatrix} X_{1\pm 1KIM}^{s's1\rho} \tag{C.5}$$

for the polarization of nucleon "1" (for nucleon "2" an extra phase factor $(-)^{s+s'}$ has to be inserted).

(ii) Two-nucleon polarization: Again it is sufficient to write down the expansions for $\bar{Y}_{IM}(x_i x_j)$ and $\bar{Z}_{IM}(x_i x_j)$

$$\bar{Y}_{IM}(xx/yy,\Theta) = -\frac{4}{1+\delta_{M0}} \sum_K \Bigg[\left(\bar{y}_{IM00}^K + \frac{1}{\sqrt{2}} \bar{y}_{IM20}^K \right) d_{M0}^K(\Theta) \mp \frac{\sqrt{3}}{2} (\bar{y}_{IM22}^K d_{M,-2}^K(\Theta) + \bar{y}_{IM2-2}^K d_{M2}^K(\Theta)) \Bigg] , \tag{C.6}$$

$$Z_{IM}(xx/yy,\Theta) = -2 \sum_K \Bigg[(z_{IM00} + z_{IM20}) d_{M-2,0}(\Theta) \mp \frac{\sqrt{3}}{2} (\bar{z}_{IM22}^K d_{M-2,-2}^K(\Theta) + \bar{z}_{IM2-2}^K d_{M+2,2}^K(\Theta)) \Bigg] , \tag{C.7}$$

$$\bar{Y}_{IM}(zz,\Theta) = -\frac{4}{1+\delta_{M0}} \sum_K (\bar{y}_{IM00}^K - \sqrt{2} \bar{y}_{IM20}^K) d_{M0}^K(\Theta) , \tag{C.8}$$

$$\bar{Z}_{IM}(zz,\Theta) = -2 \sum_K (\bar{z}_{IM00}^K - \sqrt{2} \bar{z}_{IM20}^K) d_{M-2,0}^K(\Theta) , \tag{C.9}$$

$$\bar{Y}_{IM}(xz/zx,\Theta) = -\frac{2\sqrt{3}}{1+\delta_{M0}} \sum_K [\pm(\bar{y}_{IM11}^K d_{M,-1}^K(\Theta) + \bar{y}_{IM1-1}^K d_{M1}^K(\Theta)) + \bar{y}_{IM21}^K d_{M,-1}^K(\Theta) - \bar{y}_{IM2-1}^K d_{M1}^K(\Theta)] , \tag{C.10}$$

$$\bar{Z}_{IM}(xz/zx,\Theta) = -\sqrt{3}\sum_K[\pm(\bar{z}^K_{IM11}d^K_{M-2,-1}(\Theta) + \bar{z}^K_{IM1-1}d^K_{M-2,1}(\Theta)) + \bar{z}^K_{IM21}d^K_{M-2,-1}(\Theta) - \bar{z}^K_{IM2-1}d^K_{M-2,1}(\Theta)] , \quad (C.11)$$

$$\bar{Y}_{IM}(xy/yx,\Theta) = \frac{2\sqrt{3}}{1+\delta_{M0}}\sum_K[\pm\sqrt{2}\bar{y}^K_{IM10}d^K_{M0}(\Theta) + \bar{y}^K_{IM22}d^K_{M,-2}(\Theta) - \bar{y}^K_{IM2-2}d^K_{M2}(\Theta)] , \quad (C.12)$$

$$\bar{Z}_{IM}(xy/yx,\Theta) = \sqrt{3}\sum_K[\pm\sqrt{2}\bar{z}^K_{IM10}d^K_{M-2,0}(\Theta) + \bar{z}^K_{IM22}d^K_{M-2,-2}(\Theta) - \bar{z}^K_{IM2-2}d^K_{M-2,2}(\Theta)] , \quad (C.13)$$

$$\bar{Y}_{IM}(yz/zy,\Theta) = -\frac{2\sqrt{3}}{1+\delta_{M0}}\sum_K[\pm(\bar{y}^K_{IM11}d^K_{M1}(\Theta) - \bar{y}^K_{IM1-1}d^K_{M,-1}(\Theta)) + \bar{y}^K_{IM21}d^K_{M,-1}(\Theta) + \bar{y}^K_{IM2-1}d^K_{M1}(\Theta)] , \quad (C.14)$$

$$\bar{Z}_{IM}(yz/zy,\Theta) = -\sqrt{3}\sum_K[\pm(\bar{z}^K_{IM11}d^K_{M-2,-1}(\Theta) - \bar{z}^K_{IM1-1}d^K_{M-2,1}(\Theta)) + \bar{z}^K_{IM21}d^K_{M-2,-1}(\Theta) + \bar{z}^K_{IM2-1}d^K_{M-2,1}(\Theta)] , \quad (C.15)$$

where the coefficients are defined by

$$\bar{y}/\bar{z}^K_{IMS\sigma}(1,2) = (-)^{1+M}\hat{S}\sum_{s's}(-)^{s'+s}\hat{s}'\hat{s}\begin{Bmatrix} \frac{1}{2} & \frac{1}{2} & 1 \\ \frac{1}{2} & \frac{1}{2} & 1 \\ s' & s & S \end{Bmatrix} X^{s'sS\sigma}_{1\pm1KIM} \quad (C.16)$$

and with

$$\begin{aligned} X^{s'sS\sigma}_{\lambda'\lambda KIM} = {} & \hat{I}\hat{K}^2\sum_{JK'}\hat{J}^2\hat{K}'^2\begin{pmatrix} J & I & K \\ \lambda-\lambda' & M & \lambda'-\lambda-M \end{pmatrix}\begin{pmatrix} S & K & K' \\ \sigma & -\sigma & 0 \end{pmatrix} \\ & \sum_{\substack{L\mu jl \\ L'\mu'j'l'}} (-)^{L+l}\hat{l}\hat{l}'\hat{j}\hat{j}'\begin{pmatrix} l & l' & K \\ 0 & 0 & 0 \end{pmatrix}\begin{pmatrix} L & L' & J \\ \lambda & -\lambda' & \lambda'-\lambda \end{pmatrix} \\ & \begin{Bmatrix} S & K & K' \\ s & j & l \\ s' & j' & l' \end{Bmatrix}\begin{Bmatrix} j & j' & K \\ 1 & 1 & I \\ L & L' & J \end{Bmatrix} \mathcal{O}^{L'\lambda'*}(\mu'j'l's')\mathcal{O}^{L\lambda}(\mu jls) . \end{aligned} \quad (C.17)$$

Appendix D: Derivation of the Sakita identities

These relations can be obtained from the two identities

$$\{[N_{14}, \not{q}], \not{q}\} = 0 \,, \tag{D.1}$$
$$\{N_8, \not{Q}\} = 0 \,, \tag{D.2}$$

which one verifies easily. Using

$$[N_{14}, \not{q}] = 2I(q) \tag{D.3}$$

together with (4.5.11) and the equivalence

$$\not{q} N_{14} \not{q} \mathrel{\hat{=}} q^2 N_{14} \,,$$

one obtains from the first identity (D.1)

$$\begin{aligned} 0 &= N_{14} q^2 + \not{q} N_{14} \not{q} + \frac{1}{2}\{N(\gamma,\gamma)U \cdot q + [N(q,\gamma), \not{U}], \not{q}\} \\ &\mathrel{\hat{=}} 2q^2 N_{14} + M(N_{11} + N_{12}) + 2N_4 - N_9 - 2N_6 \,, \end{aligned} \tag{D.4}$$

which is the first of Sakita's two equivalence relations

$$N_9 - 2(N_4 - N_6) \mathrel{\hat{=}} 2q^2 N_{14} + M(N_{11} + N_{12}) \,. \tag{D.5}$$

The second relation can be obtained from the second identity (D.2) using the following form of N_8

$$\begin{aligned} N_8 &= \frac{1}{2}[\not{Q}, N_{14} \not{k}] \\ &= \frac{1}{2} N_{10} \not{Q} + k \cdot Q N_{14} - I(Q) \not{k} \,. \end{aligned} \tag{D.6}$$

With the help of the following relations and equivalent forms

$$\not{Q} N_{10} \not{Q} = [\not{Q}, N_{10}] \not{Q} + N_{10} \not{Q}^2 \tag{D.7}$$
$$\frac{1}{2}[\not{Q}, N_{10}] \not{Q} = -[N_8, \not{Q}] \tag{D.8}$$
$$\mathrel{\hat{=}} 2MN_8 + 2k \cdot Q N_1 - 2k \cdot q N_2 \tag{D.9}$$
$$\{I(Q)\not{k}, \not{Q}\} \mathrel{\hat{=}} 2k \cdot q I(Q) \tag{D.10}$$

and using (4.5.13) one finds from (D.2)

$$\begin{aligned} 0 =& \frac{1}{2}\{N_{10} \not{Q}, \not{Q}\} + k \cdot Q \{N_{14}, \not{Q}\} - \{I(Q)\not{k}, \not{Q}\} \\ \mathrel{\hat{=}}& 2MN_8 + \left(Q^2 - \frac{1}{2} k \cdot Q\right) N_{10} + k \cdot Q(2N_1 - N_{13}) \\ &+ k \cdot q(2MN_{14} + N_{11} + N_{12} - 2N_2) \,. \end{aligned} \tag{D.11}$$

This is the second of Sakita's relations. We would like to mention that two other non-trivial identities

$$\{[N_{14}, \not{Q}], \not{Q}\} = 0 \tag{D.12}$$
$$\{N_9, \not{q}\} = 0 \tag{D.13}$$

similar to the ones in (D.1–2) give no new but again the above two relations.

References *

AbB 85 Abegg, R., Birchall, J., Cairns, E., Coombes, H., Davis, C.A., Davison, N.E., Green, P.W., Greeniaus, L.G., Gubler, H.P., Lee, W.P., McDonald, W.J., Miller, C.A., Moss, G.A., Plattner, G.R., Poffenberger, P.R., Roy, G., Soukup, J., Svenne, J.P., Tkachuk, R., Van Oers, W.T.H. and Zhang, Y.P.
Nucl. Instr. Meth. A234, 11 (1985)

AbS 77 Abidi, S.T.H. and Srivastava, B.K.
Can. J. Phys. 55, 943(1977)

AdA 88 Adamian, F.V., Aghababian, K.Sh., Asatrian, R.A., Bartikian, M.V., Buniatian, A.Yu., Frangulian, G.S., Galumian, P.I., Grabsky, V.H., Hairapetian, A.V., Hakopian, H.H., Karapetian, G.V., Karapetian, V.V., Kordonsky, M.S., Oganezov, R.O., Oktanian, V.K., Vartapetian, A.H., Volchinsky, V.A. and Vukolov, V.P.
Yerevan Physics Institute preprint YERPHI-1061(24)-88

AdC 70 Adler, R.J., Chertok, B.T. and Miller, H.C.
Phys. Rev. C2, 69 (1970)

AhB 73 Ahrens, J., Borchert, H., Zieger, A. and Ziegler, B.
Nucl. Instr. Meth. 108, 517 (1973)

AhE 74 Ahrens, J., Eppler, H.B., Gimm, H., Kröning, M., Riehn, P., Wäffler, H., Zieger, A. and Ziegler, B.
Phys. Lett. 52B, 49 (1974)

AlA 83 Althoff, K.H., Anton, G., Bour, D., Bock, B., Ferber, W., Gelhausen, H.W., Horikawa, N., Jahnen, Th., Kaul, O., König, W., Königsmann, K.C., Menze, D., Meyer, W., Miczaika, Th., Roderburg, E., Ruhm, W., Schilling, E.P., Schwille, W., Sundermann, D. and Wagener, K.
Z. Phys. C21, 149 (1983)

AlA 89 Althoff, K.H., Anton, G., Bock, B., Bour, D., Dostert, R., Erbs, P., Jahnen, T., Kaul, O., Kohlgart, E., Lücking, B., Menze, D., Meyer, W., Schilling, E.P., Schwille, W.J., Sundermann, D., Thiel, W. and Thiesmeyer, D.
Z. Phys. C43, 375 (1989)

AlL 89a Alvarado, J.J., Lucio M, J.L. and Pestieau, J.
Phys. Rev. C40, 531 (1989)

AlL 89b Alvarado, J.J. and Lucio M, J.L.
preprint (1989)

* A complete bibliography on deuteron photodisintegration is being prepared by the authors and will be available soon

AmB 77 Amsler, C., Brown, R.C., Bugg, D.V., Edgington, J.A., Oram, C., Axen, D., Dubois, R., Felawka, L., Jaccard, S., Keeler, R., Vávra, J., Clough, A., Gibson, D., Ludgate, G.A., Stewart, N.M., Robertson, L.P. and Richardson, J.R.
Nucl. Instr. Meth. 144, 401 (1977)

AnC 81 Anastasio, M. and Chemtob, M.
Nucl. Phys. A364, 219 (1981)

AnW 88 Annand, J., Wallace, P., Crawford, G., Hall, S., Kellie, J., McGeorge, J.C., MacGregor, D., Owens, R.O., Beck, R., Irmscher, D., Schmieden, H., Schoch, B. and Vogt, J.
Physics with MAMI A, Inst. für Kernphysik (SFB 201, University Mainz (1988)) p. 60

ArD 71 Arenhövel, H., Danos, M. and Williams, H.T.
Nucl. Phys. A162, 12 (1971)

Are 79 Arenhövel, H.
Nukleonika 24, 273 (1979)

Are 81a Arenhövel, H.
Phys. Rev. Lett. 47, 749 (1981)

Are 81b Arenhövel, H.
Lecture Notes in Physics 137, 136 (1981)

Are 81c Arenhövel, H.
Z. Phys. A302, 25 (1981)

Are 82 Arenhövel, H.
Nucl. Phys. A374, 521c (1982)

Are 83 Arenhövel, H.
Nuovo Cim. 76A, 256 (1983)

Are 85 Arenhövel, H.
Lecture Notes in Physics 234, 276 (1985)

Are 86 Arenhövel, H.
Proc. Int. School of Intermediate-Energy Nuclear Physics, Verona (Bergère, R., Costa, S., Schaerf, C., eds.), p. 55. Singapore: World Scientific 1986

Are 88a Arenhövel, H.
Few-Body Systems 4, 55 (1988)

Are 88b Arenhövel, H.
Phys. Rev. C38, 2974 (1988)

ArE 82 Arends, J., Eyink, J., Hartmann, H., Hegerath, A., Mecking, B. and Rost, H.
Nucl. Instr. Meth. 201, 361 (1982)

ArF 74 Arenhövel, H., Fabian, W. and Miller, H.G.
Phys. Lett. 52B, 303 (1974)

ArF 77a Arenhövel, H. and Fabian, W.
Nucl. Phys. A282, 397 (1977)

ArF 77b Arenhövel, H. and Fabian, W.
Nucl. Phys. A292, 429 (1977)

ArG 84 Arends, J., Gassen, H.J., Hegerath, A., Mecking, B., Nöldeke, G., Prenzel, P., Reichelt, T., Voswinkel, A. and Sapp, W.W.
Nucl. Phys. A412, 509 (1984)

ArL 86 Arenhövel, H. and Leidemann, W.
Few-Body Systems Suppl. 1, 224 (1986)

ArS 90 Arenhövel, H. and Schmitt, K.-M.
Few-Body Systems 8, 77 (1990)

ArW 72 Arenhövel, H. and Weber, H.J.
Springer Tracts in Modern Physics 65, 58 (1972)

AuR 60 Austern, N. and Rost, E.
Phys. Rev. 117, 1506 (1960)

Aus 53 Austern, N.
Phys. Rev. 92, 670 (1953)

Aus 55 Austern, N.
Phys. Rev. 100, 1522 (1955)

BaB 62 Barbiellini, G., Bologna, G., Diambrini, G. and Murtas, G.P.
Phys. Rev. Lett. 9, 396 (1962)

BaB 67 Barbiellini, G., Bernardini, C., Felicetti, F. and Murtas, G.P.
Phys. Rev. 154, 988 (1967)

BaB 86 Barannik, V.P., Bratashevskij, A.S., Gorbenko, V.G., Gushchin, V.A., Zhebrovskij, Yu.V., Zybalov, A.A., Karasev, S.P., Kolesnikov, L.Ya., Konovalov, O.G., Kulish, Yu.V., Rekalo, M.P., Rubashkin, A.L., Sorokin, P.V., Storozhenko, Yu.O. and Tenishev, A.Eh.
Nucl. Phys. A451, 751 (1986)

BaC 52 Barnes, C.A., Carver, J.H., Stafford, G.H. and Wilkinson, D.H.
Phys. Rev. 86, 359 (1952)

BaC 73 Baglin, J.E.E., Carr, R.W., Bentz, E.J., Jr. and Wu, C.P.
Nucl. Phys. A201, 593 (1973)

BaE 83 Baba, K., Endo, I., Fukuma, H., Inoue, K., Kawamoto, T., Ohsugi, T., Sumi, Y., Takeshita, T., Uehara, S., Yamo, Y. and Maki, T.
Phys. Rev. C28, 286 (1983)

BaG 83 Barannik, V.P., Gorbenko, V.G., Gushchin, V.A., Zhebrovskii, Yu.V., Kolesnikov, L.Ya., Kulish, Yu.V., Rubashkin, A.L. and Sorokin, P.V.
Yad. Fiz. 38, 1108 (1983) (Sov. J. Nucl. Phys. 38, 667 (1983))

Bai 79 Baier, H.
Fortschr. Phys. 27, 209 (1979)

BeB 80 Beil, H. and Bergère, R.
Report CEA-N-2144, CEN Saclay (1980)

BeD 63 Bertozzi, W., Demos, P.T., Kowalski, S., Sargent, C.P., Turchinetz, W., Fullwood, R. and Russell, J.
Phys. Rev. Lett. 10, 106 (1963)

BeH 34 Bethe, H. and Heitler, W.
Proc. Roy. Soc. A146, 83 (1934)

BeI 86 Bernabei, R., Incicchitti, A., Mattioli, M., Picozza, P., Prosperi, D., Casano, L., d'Angelo, S., De Pascale, M.P., Schaerf, C., Giordano, G., Matone, G., Frullani, S. and Girolami, B.
Phys. Rev. Lett. 57, 1542 (1986)

BeL 50 Bethe, H.A. and Longmire, C.
Phys. Rev. 77, 647 (1950)

BeP 35 Bethe, H.A. and Peierls, R.
Proc. Roy. Soc. A148, 146 (1935)

Ber 54 Berger, J.M.
Phys. Rev. 94, 1698 (1954)

BiB 51 Bishop, G.R., Beghian, L.E. and Halban, H.
Phys. Rev. 83, 1052 L (1951)

BiB 88 Birenbaum, Y., Berant, Z., Wolf, A., Kahane, S. and Moreh, R.
Phys. Rev. Lett. 61, 810 (1988)

BiC 50 Bishop, G.R., Collie, C.H., Halban, H., Hedgran, S., Siegbahn, K., du Toit, S. and Wilson, R.
Phys. Rev. 80, 211 (1950) [Erratum: ibid. 81, 644 (1951)]

BiH 51 Bishop, G.R., Halban, H., Shaw, P.F.D. and Wilson, R.
Phys. Rev. 81, 219 (1951)

BiK 85 Birenbaum, Y., Kahane, S. and Moreh, R.
Phys. Rev. C32, 1825 (1985)

BiW 57 Bishop, G.R. and Wilson, R.
Handbuch der Physik (Flügge, S., ed.) Vol. XLII, p. 309. Berlin - Göttingen - Heidelberg: Springer 1957

BjD 64 Bjorken, J.D. and Drell, S.D.
Relativistic Quantum Mechanics. New York: McGraw-Hill 1964

BlB 52 Blatt, J.M. and Biedenharn, L.C.
Phys. Rev. 86, 399 (1952)

BlS 66 Blankenbecler, R. and Sugar, R.
Phys. Rev. 142, 1051 (1966)

BoB 79 Bosman, M., Bol, A., Gilot, J.F., Leleux, P., Lipnik, P. and Macq, P.
Phys. Lett. 82B, 212 (1979)

BoC 63 Bosco, B., Ciocchetti, G. and Molinari, A.
Nuovo Cim. 28, 1427 (1963)

BoL 87 Bohigas, O. and Lipparini, E.
Phys. Rev. C36, 2150 (1987)

BoR 79 Botzian, R., Rustgi, M.L. and Torruella, A.J.
Nuovo Cim. 52A, 309 (1979)

BrD 80 Bratashevskii, A.S., Derebchinskii, A.I., Zybalov, A.A., Konovalov, O.G., Sorokin, P.V. and Tenishev, A.E.
Yad. Fiz. 31, 860 (1980) (Sov. J. Nucl. Phys. 31, 444 (1980))

Bre 73 Breit, G.
Proc. Int. Conference on Photonuclear Reactions and Applications (Berman, B.L., ed.), p. 323. Livermore: University of California 1973

BrG 72 Bryan, R.A. and Gersten, A.
Phys. Rev. D6, 341 (1972)

BrG 86 Bratashevskij, A.S., Gorbenko, V.G., Gushchin, V.A., Zhebrovskii, Yu.V., Zybalov, A.A., Karasev, S.P., Kolesnikov, L.Ya., Konovalov, O.G., Rubashkin, A.L., Sorokin, P.V., Storozhenko, Yu.O. and Tenishev, A.E.
Yad. Fiz. 43, 785 (1986) (Sov. J. Nucl. Phys. 43, 499 (1986))

BrH 83 Brodsky, S.J. and Hiller, J.R.
Phys. Rev. C18, 475 (1983)

BrS 69 Bryan, R.A. and Scott, B.L.
Phys. Rev. 177, 1435 (1969)

BuL 85 Buchmann, A., Leidemann, W. and Arenhövel, H.
Nucl. Phys. A443, 726 (1985)

CaC 78 Caloi, R., Casano, L., Federici, L., Frullani, S., Giordano, G., Matone, G., Mattioli, M., Pasquariello, G., Picozza, P., Prosperi, D. and Schaerf, C.
Report LNF-78/34(R) 1978

CaD 80 Capitani, G.P., De Sanctis, E., Di Giacomo, P., Guaraldo, C., Lucherini, V., Polli, E., Reolon, A.R. and Scrimaglio, R.
Nucl. Instr. Meth. 178, 61 (1980)

CaD 83 Capitani, G.P., De Sanctis, E., Guaraldo, C., Di Giacomo, P., Lucherini, V., Polli, E., Reolon, A.R., Scrimaglio, R., Anghinolfi, M., Corvisiero, P., Ricco, G., Sanzone, M. and Zucchiatti, A.
Nucl. Instr. Meth. 216, 307 (1983)

CaD 84 Cameron, J.M., Davis, C.A., Fielding, H., Kitching, P., Soukup, J., Uegaki, J., Wesick, J., Wilson, H.S., Abegg, R., Hutcheon, D.A., Miller, C.A., Stetz, A.W., Shin, Y.M., Stevenson, N. and van Heerden, I.J.
Phys. Lett. 137B, 315 (1984)

CaD 86 Cameron, J.M., Davis, C.A., Fielding, H., Kitching, P., Pasos, J., Soukup, J., Uegaki, J., Wesick, J., Wilson, H.S., Abegg, R., Hutcheon, D.A., Miller, C.A., Stetz, A.W. and van Heerden, I.J.
Nucl. Phys. A458, 637 (1986)

Cam 84 Cameron, J.
Can. J. Phys. 62, 1019 (1984)

CaM 81 Cambi, A., Mosconi, B. and Ricci, P.
Phys. Rev. C23, 992 (1981)

CaM 82a Cambi, A., Mosconi, B. and Ricci, P.
Phys. Rev. C26, 2358 (1982)

CaM 82b Cambi, A., Mosconi, B. and Ricci, P.
Phys. Rev. Lett. 48, 462 (1982)

CaM 84 Cambi, A., Mosconi, B. and Ricci, P.
Proc. Workshop on Perspectives in Nuclear Physics at Intermediate Energies, Trieste 1983 (Boffi, S., Ciofi degli Atti, C. and Giannini, M.M., eds.), p. 139. Singapore: World Scientific 1984;
J. Phys. G 10, L11 (1984), and private communication

CaM 85 Cambi, A., Mosconi, B. and Ricci, P.
J. Phys. G 11, 897 (1985)

CeI 61 Cerineo, M., Ilakovac, K., Slaus, I. and Tomas, P.
Phys. Rev. 124, 1947 (1961)

Che 79 Chemtob, M.
Mesons in Nuclei II (Rho, M. and Wilkinson, D.H., eds.), p. 495. Amsterdam: North-Holland 1979

ChG 34 Chadwick, J. and Goldhaber, M.
Nature 134, 237 (1934)

ChR 71 Chemtob, M. and Rho, M.
Nucl. Phys. A163, 1 (1971)

ChS 66 Ching, R. and Schaerf, C.
Phys. Rev. 141, 1320 (1966)

Col 51 Colgate, S.A.
Phys. Rev. 83, 1262 L (1951)

CoM 73 Colocci, M., Mosconi, B. and Ricci, P.
Phys. Lett. 45B, 224 (1973)

CoW 65 Cox, A.E., Wynchank, S.A.R. and Collie, C.H.
Nucl. Phys. 74, 497 (1965)

CrT 78 Craver, B.A., Tubis, A. and Kim, Y.E.
Phys. Rev. 18, 1559 (1978)

CrZ 82 Crepinsek, L. and Zingl, H.F.K.
Acta Phys. Austr. 54, 151 (1982)

CzS 57 Czyz, W. and Sawicki, J.
Nuovo Cim. 5, 45 (1957)

CzS 58a Czyz, W. and Sawicki, J.
Phys. Rev. 110, 900 (1958)

CzS 58b Czyz, W. and Sawicki, J.
Nucl. Phys. 8, 621 (1958)

DaV 67 Davey, P.O. and Valk, H.S.
Phys. Rev. 156, 1039 (1967)

DBF 81 Del Bianco, W., Federici, L., Giordano, G., Matone, G., Pasquariello, G., Picozza, P., Caloi, R., Casano, L., De Pascale, M.P., Ingrosso, L., Mattioli, M., Poldi, E., Schaerf, C., Pelfer, P., Prosperi, D., Frullani, S., Girolami, B. and Jeremie, H.
Phys. Rev. Lett. 47, 1118 (1981)

DBJ 80 Del Bianco, W., Jeremie, H., Irshad, M. and Kajrys, G.
Nucl. Phys. A343, 121 (1980)

DBK 80 Del Bianco, W. and Kajrys, G.
Can. J. Phys. 58, 1496 (1980)

DGr 88 De Graeve, A.
Ph.D. thesis, University of Gent, Gent (1988)

DGZ 89 De Graeve, A., Zieger, A., Van de Vyver, R., Ferdinande, H., Van Hoorebeke, L., Ryckbosch, D., Van Otten, P. and Ziegler, B.
Phys. Lett. B227, 321 (1989)

Dia 68 Diambrini, G.
Rev. Mod. Phys. 40, 611 (1968)

Don 62a Donnachie, A.
Nucl. Phys. 32, 637 (1962)

Don 62b Donnachie, A.
Nucl. Phys. 37, 594 (1962)

DoK 76 Dougan, P., Kivikas, T., Lugner, K., Ramsey, V. and Stiefler, W.
Z. Phys. A276, 55 (1976)

DoO 64a Donnachie, A. and O'Donnell, P.J.
Nucl. Phys. 53, 128 (1964)

DoO 64b Donnachie, A. and O'Donnell, P.J.
Nucl. Phys. 59, 385 (1964)

DoR 77 Dougan, P., Ramsay, V. and Stiefler, W.
Z. Phys. A280, 341 (1977)

DPG 82 De Pascale, M.P., Giordano, G., Matone, G., Picozza, P., Azario, L., Caloi, R., Casano, L., Ingrosso, L., Mattioli, M., Poldi, E., Prosperi, D. and Schaerf, C.
Phys. Lett. 119B, 30 (1982)

DPG 85 De Pascale, M.P., Giordano, G., Matone, G., Babusci, D., Bernabei, R., Bilaniuk, O.M., Casano, L., d'Angelo, S., Mattioli, M., Picozza, P., Prosperi, D., Schaerf, C., Frullani, S. and Girolami, B.
Phys. Rev. C32, 1830 (1985)

DrG 76 Dressler, E.J. and Gross, F.
Nucl. Phys. A262, 516 (1976)

Dro 76 Drooks, L.J.
Ph.D. thesis, Yale University, New Haven (1976)

DSA 86 De Sanctis, E., Anghinolfi, M., Capitani, G.P., Corvisiero, P., Di Giacomo, P., Guaraldo, C., Lucherini, V., Polli, E., Reolon, A.R., Ricco, G., Sanzone, M. and Zucchiatti, A.
Phys. Rev. C34, 413 (1986)

DSC 59 de Swart, J.J., Czyz, W. and Sawicki, J.
Phys. Rev. Lett. 2, 51L (1959)

DSC 85 De Sanctis, E., Capitani, G.P., Di Giacomo, P., Guaraldo, C., Lucherini, V., Polli, E., Reolon, A.R., Scrimaglio, R., Anghinolfi, M., Corvisiero, P., Ricco, G., Sanzone, M. and Zucchiatti, A.
Phys. Rev. Lett. 54, 1639 (1985)

DSM 58 de Swart, J.J. and Marshak, R.E.
Phys. Rev. 111, 272 (1958)

DSM 59 de Swart, J.J. and Marshak, R.E.
Physica 25, 1001 (1959)

DSw 59 de Swart, J.J.
Physica 25, 233 (1959)

DTR 75 De Tourreil, R., Rouben, B. and Sprung, D.W.L.
Nucl. Phys. A242, 445 (1975)

DTS 73 De Tourreil, R. and Sprung, D.W.L.
Nucl. Phys. A201, 193 (1973)

DuK 83 Dumbrajs, O., Koch, R., Pilkuhn, H., Oades, G.C., Behrens, H., de Swart, J.J. and Kroll, P.
Nucl. Phys. B216, 277 (1983)

DuL 85 Dupont, C., Leleux, P., Lipnik, P., Macq, P. and Ninane, A.
Nucl. Phys. A445, 13 (1985)

Edw 89 Edwards, G.W.R.
Ph.D. thesis, University of Alberta, Edmonton (1989)

FaA 76 Fabian, W. and Arenhövel, H.
Nucl. Phys. A258, 461 (1976)

FaR 59 Fano, U. and Racah, G.
Irreducible Tensorial Sets. New York: Academic Press 1959

FeG 80 Federici, L., Giordano, G., Matone, G., Pasquariello, G., Picozza, P.G., Caloi, R., Casano, L., De Pascale, M.P., Mattioli, M., Poldi, E., Schaerf, C., Vanni, M., Pelfer, P., Prosperi, D., Frullani, S. and Girolami, B.
Nuovo Cim. 59B, 247 (1980)

Fel 55 Feld, B.T.
Nuovo Cim., Ser. X., Suppl. II, 145 (1955)

Fer 35 Fermi, E.
Phys. Rev. 48, 570 (1935)

FeS 51 Feshbach, H. and Schwinger, J.
Phys. Rev. 84, 194 (1951)

FiD 89 Fink, G., Doll, P., Hauber, S., Haupenthal, M., Klages, H.O., Schieler, H., Smend, F. and Wicke, G.D.
Preprint KfZ Karlsruhe (1989)

Fra 58 Frantz, L.M.
Phys. Rev. 111, 346 (1958)

FrF 83 Friar, J.L., Fallieros, S., Tomusiak, E.L., Skopik, D. and Fuller, E.G.
Phys. Rev. C27, 1364 (1983)

FrF 84 Friar, J.L. and Fallieros, S.
Phys. Rev. C29, 1645 (1984)

FrG 84 Friar, J.L., Gibson, B.F. and Payne, G.L.
Phys. Rev. C30, 441 (1984)

Fri 77 Friar, J.L.
Ann. Phys. (NY) 104, 380 (1977)

Fri 79 Friar, J.L.
Mesons in Nuclei II (Rho, M. and Wilkinson, D.H., eds.), p. 595. Amsterdam: North-Holland 1979

GaH 73 Gari, M. and Huffmann, A.H.
Phys. Rev. C7, 994 (1973)

GaH 76 Gari, M. and Hyuga, H.
Z. Phys. A227, 291 (1976)

GaH 78 Gari, M., Hebach, H., Sommer, B. and Zabolitzky, J.G.
Phys. Rev. Lett. 41, 1288 (1978)

GaH 81 Gari, M. and Hebach, H.
Phys. Rep. 72, 1 (1981)

GaS 78 Gari, M. and Sommer, B.
Phys. Rev. Lett. 41, 22 (1978)

Gen 49 Genevese, F.
Phys. Rev. 76, 1288 (1949)

Geo 68 George, D.J.
Phys. Rev. 167, 1357 (1968)

GiB 81 Gilot, J.F., Bol, A., Leleux, P., Lipnik, P. and Macq, P.
Phys. Rev. Lett. 47, 304 (1981)

GoL 81 Govaerts, J., Lucio, J.L., Martinez, A. and Pestieau, J.
Nucl. Phys. A365, 477 (1981)

GoL 82 Goulard, G. and Lorazo, B.
Can. J. Phys. 60, 162 (1982)

GoZ 79 Gorbenko, V.G., Zhebrovskii, Yu.V., Kolesnikov, L.Ya., Rubashkin, A.L. and Sorokin, P.
Pis'ma Zh. Eksp. Teor. Fiz. 30, 130 (1979) (Sov. Phys. JETP Lett. 30, 118 (1979))

GoZ 82 Gorbenko, V.G., Zhebrovskii, Yu.V., Kolesnikov, L.Ya., Rubashkin, A.L. and Sorokin, P.
Nucl. Phys. A381, 330 (1982)

Gre 76 Green, A.M.
Rep. Prog. Phys. 39, 1109 (1976)

GrH 45 Graham, G.A.R. and Halban, H.
Rev. Mod. Phys. 17, 297 (1945)

Gri 89 Grimm, H.
Ph.D. thesis, University Mainz, Mainz (1989)

GrS 79 Green, A.M. and Sainio, M.
J. Phys. G5, 503 (1979)

GrS 82a Green, A.M. and Sainio, M.
J. Phys. G8, 1337 (1982)

GrS 82b Green, A.M. and Strueve, W.
Phys. Lett. 112B, 10 (1982); [Erratum: ibid. 112B, 495 (1982)]

GrW 83 Greben, J.M. and Woloshyn, R.M.
J. Phys. G9, 643 (1983)

Had 79 Hadjimichael, E.
Phys. Lett. 85B, 17 (1979)

HaJ 62 Hamada, T. and Johnston, I.D.
Nucl. Phys. 34, 3892 (1962)

HaR 87 Hadjimichael, E., Rustgi, M.L. and Pandey, L.N.
Phys. Rev. C36, 44 (1987)

Has 68 Hasselmann, D.
DESY preprint 68/38 (1968)

HaS 80 Hadjimichael, E. and Saylor, D.P.
Phys. Rev. Lett. 45, 1776 (1980)

HaW 49a Hamermesh, B. and Wattenberg, A.
Phys. Rev. 75, 1290 (1949)

HaW 49b Hamermesh, B. and Wattenberg, A.
Phys. Rev. 76, 1408 L (1949)

Hei 54 Heitler, W.
The Quantum Theory of Radiation. London: Oxford University Press 1954

HoS 83 Holt, J.R., Stephenson, K. and Specht, J.R.
Phys. Rev. Lett. 50, 577 (1983)

Hou 50 Hough, P.V.C.
Phys. Rev. 80, 1069 (1950)

Hsi 56a Hsieh, S.H.
Prog. Theor. Phys. 16, 68L (1956)

Hsi 56b Hsieh, S.H.
Nuovo Cim. 4, 138 (1956)

Hsi 59a Hsieh, S.H.
Prog. Theor. Phys. 21, 185 (1959)

Hsi 59b Hsieh, S.H.
Prog. Theor. Phys. 21, 211L (1959)

Hsi 59c Hsieh, S.H.
Prog. Theor. Phys. 21, 585 (1959)

HuC 87 Hugi, M., Cameron, J.H., Ahmad, M., Collot, J., Gaillard, G., Wesick, J.S., Edwards, G.W.R., Fielding, H., Hutcheon, D.H., Abegg, R., Miller, C.A., Kitching, P., Davison, N.E., Stevenson, N.R., and Van Heerden, I.J.
Nucl. Phys. A472, 701 (1987)

HuM 49 Hu, T.M. and Massey, H.S.W.
Proc. Roy. Soc. A196, 135 (1949)

Hun 80 Huneke, H.
Diploma thesis, University Bonn, IR-80-24, Bonn (1980)

HuS 57 Hulthèn, L. and Sugawara, M.
Handbuch der Physik (Flügge S., ed.) Vol. XXXIX, p. 95. Berlin - Göttingen - Heidelberg: Springer 1957

Hut 84 Hutcheon, D.A.
Can. J. Phys. 62, 1114 (1984)

HuZ 76 Hughes, R.J., Zieger, A., Wäffler, H. and Ziegler, B.
Nucl. Phys. A267, 329 (1976)

Hwa 84 Hwang, W.Y.P.
Can. J. Phys. 62, 1072 (1984)

HwL 83 Hwang, W.Y.P., Londergan, J.T. and Walker, G.E.
Ann. Phys. (N.Y.) 149, 335 (1983)

HwM 80 Hwang, W.Y.P. and Miller, G.A.
Phys. Rev. C22, 968 (1980); [Erratum: ibid. C24, 325 (1981)]

HwW 85 Hwang, W.Y.P. and Walker, G.E.
Ann. Phys. 159, 118 (1985)

IkA 79 Ikeda, H., Arai, I., Fujii, H., Fujii, T., Iwasaki, H., Kajiura, N., Kamae, T., Nakamura, K., Sumiyoshi, T. and Takeda, H.
Phys. Rev. Lett. 42, 1321 (1979)

IsK 82 Ishii, T., Kato, S., Okuno, H., Ukai, K., Ohashi, Y., Awaji, N., Hayashii, H., Horikawa, N., Kimura, K., Miyamoto, A., Nakanishi, T., Okumi, S., Ozaki, H., Sugiyama, A., Suzaki, R., Tauchi, T., Mori, K. and Takeuchi, Y.
Phys. Lett. 110B, 441 (1982)

JaW 81 Jaus, W. and Woolcock, W.S.
Nucl. Phys. A365, 477 (1981)

JaW 84a Jaus, W. and Woolcock, W.S.
Nucl. Phys. A431, 669 (1984)

JaW 84b Jaus, W. and Woolcock, W.S.
Helv. Phys. Acta 57, 644 (1984)

JaW 87a Jaus, W. and Woolcock, W.S.
Nuovo Cim. 97A, 103 (1987)

JaW 87b Jaus, W. and Woolcock, W.S.
Nucl. Phys. A473, 667 (1987)

JeJ 65 Jewell, R.W., John, W., Sherwood, J.E. and White, D.H.
Phys. Rev. 138B, 71 (1965)

KaA 77 Kamae, T., Arai, I., Fujii, T., Ikeda, H., Kajiura, N., Kawabata, S., Nakamura, K., Ogawa, K., Takeda, H. and Watase, Y.
Phys. Rev. Lett. 38, 468 (1977)

KaL 66 Kaschluhn, F. and Lewin, K.
Nucl. Phys. 87, 73 (1966)

KaL 72 Kaschluhn, F. and Lewin, K.
Nucl. Phys. B49, 525 (1972)

KaL 77 Kaschluhn, F. and Lewin, K.
Annalen Phys. 34, 67 (1977)

Kan 86 Kang, Y.
Diploma thesis, University Bonn, IR-86-20, Bonn (1986)

Kaw 58 Kawaguchi, M.
Phys. Rev. 111, 1314 (1958)

KeM 78 Kermode, M.W. and McKerrell, A.
Lecture Notes in Physics 82, 63 (1978)

KoM 59 Koch, H.W. and Motz, J.W.
Rev. Mod. Phys. 31, 920 (1959)

KoM 69 Kose, R., Martin, B., Runkel, R., Wahlen, H. and Kissler, K.H.
Z. Phys. 220, 305 (1969)

KrM 60 Kramer, G. and Müller, D.
Z. Phys. 158, 204 (1960)

KrS 88 Krause, K.H., Sobolewski, J., Ahrens, J., Henneberg, J.M. and Ziegler, B.
Physics with MAMI A, Inst. für Kernphysik/SFB 201, University Mainz (1988), p. 85

Lag 78 Laget, J.M.
Nucl. Phys. A312, 265 (1978)

Lag 84 Laget, J.M.
Can. J. Phys. 62, 1046 (1984)

LaH 62 Lassila, K.E., Hull, Jr., M.H., Rupple, H.M., Mac Donald, F.A. and Breit, G.
Phys. Rev. 126, 881 (1962)

LaL 80 Lacombe, M., Loiseau, B., Richard, J.M., Vinh Mau, R., Côté, J., Pirès, P. and de Tourreil, R.
Phys. Rev. C21, 861 (1980)

Las 49 Lassen, N.O.
Phys. Rev. 75, 1099 L (1949)

LeA 83 Leidemann, W. and Arenhövel, H.
Nucl. Phys. A393, 385 (1983)

LeA 84 Leidemann, W. and Arenhövel, H.
Can. J. Phys. 62, 1036 (1984)

LeA 87 Leidemann, W. and Arenhövel, H.
Nucl. Phys. A465, 573 (1987)

LeA 89 Levi Sandri, P., Anghinolfi, M., Bianchi, N., Capitani, G.P., Corvisiero, P., De Sanctis, E., Guaraldo, C., Lucherini, V., Muccifora, V., Polli, E., Reolon, A.R., Ricco, G., Rossi, P., Sanzone, M. and Taiuti, M.
Phys. Rev. C39, 1701 (1989)

LeB 50 Levinger, J.S. and Bethe, H.A.
Phys. Rev. 78, 115 (1950)

LeD 87 Levi Sandri, P., De Sanctis, E., Anghinolfi, M., Bianchi, N., Capitani, G.P., Corvisiero, P., Guaraldo, C., Lucherini, V., Muccifora, V., Polli, E., Reolon, A.R., Ricco, G., Rossi, P., Sanzone, M. and Taiuti, M.
Phys. Rev. Lett. 59, 2543 (1987)

Lee 84 Lee, T.-S.H.
Phys. Rev. C29, 195 (1984)

Lee 88 Lee, T.-S.H.
preprint PHY-5253-TH-88 (1988)

Lei 80 Leidemann, W.
Diploma thesis, University Mainz, Mainz (1980)

LeR 64a LeBellac, M., Renard, F.M. and Trân Thanh Vân, J.
Nuovo Cim. 33, 594 (1964)

LeR 64b LeBellac, M., Renard, F.M. and Trân Thanh Vân, J.
Nuovo Cim. 34, 450 (1964)

LeS 81 Leicht, R., Schelhaas, K.P., Hammen, M., Ahrens, J. and Ziegler, B.
Nucl. Instr. Meth. 179, 131 (1981)

Lev 49 Levinger, J.S.
Phys. Rev. 76, 699L (1949)

Lev 55 Levinger, J.S.
Phys. Rev. 97, 970 (1955)

Lev 60 Levinger, J.S.
Nuclear Photodisintegration. London: Oxford University Press 1960

Lev 88 Levchook, M.I.
Contr. 17th INS Int. Symposium on Nuclear Physics at Intermediate Energy, Tokyo 1988, Abstracts (Inst. for Nuclear Study, Univ. of Tokyo, 1988) p. 64

LiL 68 Liu, F.F., Lundquist, D.E. and Wiik, B.H.
Phys. Rev. 165, 1478 (1968)

Lip 77 Lipparini, E.
Nuovo Cim. 37A, 279 (1977)

Liu 64 Liu, F.F.
Phys. Lett. 11, 306 (1964)

Liu 65 Liu, F.F.
Phys. Rev. 138, B1443 (1965)

Loc 84 Locher, M.P.
Nucl. Phys. A416, 243c (1984)

Lom 77 Lomon, E.L.
Phys. Lett. 68B, 419 (1977)

LoS 86 Locher, M.P., Sainio, M. and Švarc, A.
Adv. Phys. 17, 47 (1986)

LuM 85 Lucio, J.L., Martinez, A. and Pestieau, J.
Nuovo Cim. 89A, 292 (1985)

LuR 68 Lucas, J.G. and Rustgi, M.L.
Nucl. Phys. A112, 503 (1968)

Lvo 87 L'vov, A.I.
Prob. Atom. Sc. Tech., Ser.: Gen. and Nucl. Phys. 38, 93 (1987) (in Russian)

MaB 54 Marin, P., Bishop, G.R. and Halban, H.
Proc. Phys. Soc. 67A, 1113 (1954)

Mac 89 Machleidt, R.
Adv. Nucl. Phys. 19, 189 (1989)

Mad 71 Madison Convention, Proc. Third Int. Symposium on Polarization Phenomena in Nuclear Reactions (Barshall, H.H. and Haeberli, W., eds.), Madison: University of Wisconsin Press, 1971, p. XXV

MaG 50 Marshall, J.F. and Guth, E.
Phys. Rev. 78, 738 (1950)

MaH 87 Machleidt, R., Holinde, K. and Elster, Ch.
Phys. Rep. 149, 1 (1987)

Mak 84 Makarov, M.M.
Fiz. Elem. Chastits At. Yadra 15, 941 (1984) (Sov. J. Part. Nucl. 15, 419 (1984))

Mar 88 Maruyama, K.
Nucl. Phys. A478, 523c (1988)

MaS 68 Mancini, E. and Sanzone, M.
Nucl. Instr. Meth. 66, 87 (1968)

Mat 60 Matsumoto, M.
Prog. Theor. Phys. 23, 597 (1960)

Mat 63 Matsumoto, M.
Phys. Rev. 129, 1334 (1963)

Mat 82 Mathiot, J.F.
Phys. Lett. 115B, 174 (1982)

Mat 89 Mathiot, J.
Phys. Rep. 173, 63 (1989)

Mat 90 Matthews, J.L.
Private communication

MaV 61 Martin, A. and Vinh Mau, R.
Nuovo Cim. 20, 194L (1961)

McT 83 McKerrell, A., McTavish, J.P., Kermode, M.W. and Huby, R.
J. Phys. G9, 275 (1983)

MeH 84 Meyer, H.O., Hall, J.R., Hugi, M., Karwowski, H.J., Pollok, R.E. and Schwandt, R.
Phys. Rev. Lett. 52, 1759 (1984)

MeH 85 Meyer, H.O., Hall, J.R., Hugi, M., Karwowski, H.J., Pollok, R.E. and Schwandt, R.
Phys. Rev. C31, 309 (1985)

Mei 49 Meiners, E.P., Jr.
Phys. Rev. 76, 259 (1949)

Mey 85a Meyer, W.
Lecture Notes in Physics 234, 413 (1985)

Mey 85b Meyer, W.
J. Phys. Colloq. (France) 46, 455 (1985)

MMC 55 McMurray, W.R. and Collie, C.H.
Proc. Phys. Soc. 68A, 181 (1955)

MoK 89 Moreh, R., Kennett, T.J. and Prestwich, W.V.
Phys. Rev. C39, 1247 (1989) [Erratum: ibid. C40, 1548 (1989)]

MoN 87 Mostovoy, M.V., Nikolenko, D.M., Ospanov, K.T., Popov, S.G., Rachek, I.A., Shatnov, Yu.M., Toporkov, D.K., Tsentalovich, E.P., Woitsekhowski, B.B., Zelevinsky, V.G., Naumenko, G.A. and Stibunov, V.N.
Phys. Lett. B188, 181 (1987)

MoR 87 Mosconi, B. and Ricci, P.
Phys. Rev. C36, 60 (1987)

MoR 88 Mosconi, B., Ricci, P. and Cambi, A.
Phys. Rev. C38, 2976 (1988)

MuR 67 Mukherjee, S.N., Roy, B.B. and Rustgi, M.L.
Phys. Lett. 24B, 548 (1967)

MyG 61 Myers, H., Gomez, R., Guinier, D. and Tollestrup, A.V.
Phys. Rev. 121, 630 (1961)

NaF 72 Nath, R., Firk, F.W.K. and Schultz, H.L.
Nucl. Phys. A194, 49 (1972)

NaF 88 Napolitano, J., Freedman, S.J., Geesaman, D.F., Gilman, R., Green, M.C., Holt, R.J., Jackson, H.E., Kowalczyk, R., Marchand, C., Nelson, J., Zeidman, B., Beck, D., Boyd, G., Collins, D., Filippone, B.W., Jourdan, J., McKeown, R.D., Milner, R., Potterveld, D., Walker, R., Woodward, C., Segel, R.E., Tung, T.-Y., Bosted, P.E., Kinney, E.R., Meziani, Z.-E. and Minehart, R.
Phys. Rev. Lett. 61, 2530 (1988)

NaK 86 Nagornyi, S.I., Kasatkin, Yu.A., Inopin, E.V. and Kirichenko, I.K.
Yad. Fiz. 44, 1171 (1986) (Sov. J. Nucl. Phys. 44, 760 (1986))

NiB 59 Nicholson, A.F. and Brown, G.E.
Proc. Phys. Soc. 73A, 221 (1959)

NiD 84 Ninane, A., Dupont, C., Leleux, P., Lipnik, P. and Macq, P.
Can. J. Phys. 62, 1104 (1984)

NiD 87 Ninane, A., Dupont, C., Leleux, P., Lipnik, P. and Macq, P.
Phys. Rev. C35, 402 (1987)

Noy 65 Noyes, H.P.
Nucl. Phys. 74, 508 (1965)

ODo 63 O'Donnell, P.J.
Nucl. Phys. 49, 475 (1963)

OgK 80 Ogawa, K., Kamae, T. and Nakamura, K.
Nucl. Phys. A340, 451 (1980)

Oht 89 Ohta, K.
Nucl. Phys. A495, 564 (1989)

OlM 59 Olsen, H. and Maximon, L.C.
Phys. Rev. 114, 887 (1959)

Owe 90 Owens, R.O.
Private communication

Par 64 Partovi, F.
Ann. Phys. (N.Y.) 27, 79 (1964)

Par 87 Partovi, F.
Phys. Rev. C36, 491 (1987)

PaR 85a Pandey, L.N. and Rustgi, M.L.
Phys. Rev. C32, 1191 (1985)

PaR 85b Pandey, L.N. and Rustgi, M.L.
Phys. Rev. C32, 1842 (1985)

PeK 60 Pearlstein, L.D. and Klein, A.
Phys. Rev. 118, 193 (1960)

RaS 71 Raghavan, S.S. and Shrivastava, B.K.
Can. J. Phys. 49, 2211 (1971)

Rei 68 Reid, R.V., Jr.
Ann. Phys. (N.Y.) 50, 411 (1968)

ReM 67 Reitan, A. and Myhrer, F.
Nucl. Phys. B3, 130 (1967)

RiB 72 Riska, D.O. and Brown, G.E.
Phys. Lett. 38B, 193 (1972)

Ris 85 Riska, D.O.
Physica Scripta 31, 471 (1985)

Ris 89 Riska, D.O.
Phys. Rep. 181, 207 (1989)

Rob 74 Robson, B.A.
The Theory of Polarization Phenomena. Oxford: Clarendon Press 1974

RoD 89a Rossi, P., De Sanctis, E., Levi Sandri, P., Bianchi, N., Guaraldo, C., Lucherini, V., Muccifora, V., Polli, E., Reolon, A.R. and Urciuoli, G.M.
Report LNF-89/017(P) (1989)

RoD 89b Rossi, P., De Sanctis, E., Levi Sandri, P., Bianchi, N., Guaraldo, C., Lucherini, V., Muccifora, V., Polli, E., Reolon, A.R. and Urciuoli, G.M.
Phys. Rev. C40, 2412 (1989)

RoG 47 Rose, M.E. and Goertzel, G.
Phys. Rev. 72, 749 (1947)

Ros 56 Rosentsveig, L.N.
Zh. Eksp. Teor. Fiz. 31, 166L (1956) (Sov. Phys. JETP 4, 280 (1956))

Ros 57 Rose, M.E.
Elementary Theory of Angular Momentum. New York: Wiley 1957

RuL 57 Rustgi, M.L. and Levinger, J.S.
Prog. Theor. Phys. 18, 100L (1957)

RuP 89 Rustgi, M.L. and Pandey, L.N.
Phys. Rev. C40, 1581 (1989)

RuR 77 Rustgi, M.L., Rustgi, O.P. and Sandhu, T.S.
Can. J. Phys. 55, 158 (1977)

RuS 77 Rustgi, M.L., Sandhu, T.S. and Rustgi, O.P.
Phys. Lett. 70B, 145 (1977)

RuV 83a Rustgi, M.L., Vyas, R. and Chopra, M.
Phys. Rev. Lett. 50, 236 (1983)

RuV 83b Rustgi, M.L. and Vyas, R.
Phys. Lett. 121B, 365 (1983)

RuZ 60 Rustgi, M.L., Zernik, W., Breit, G. and Andrews, D.J.
Phys. Rev. 120, 1881 (1960)

SaG 62 Sakita, B. and Goebel, C.J.
Phys. Rev. 127, 1787 (1962)

Sak 62 Sakita, B.
Phys. Rev. 127, 1800 (1962)

SaL 83 Sandorfi, A.M., Le Vine, M.J., Thorn, C.E., Giordano, G., Matone, G. and Schaerf, C.
IEEE Trans. Nucl. Sc. 30, 3083 (1983)

San 86 Sanzone, M.
Proc. Int. School of Intermediate Energy Nuclear Physics, Verona (Bergère, R., Costa, S., Schaerf, C., eds.), p. 78. Singapore: World Scientific 1986

San 88 Sanzone-Arenhövel, M.
Proc. Workshop on Heavy-Quark Factory and Nuclear Physics Facility with Superconducting Linacs (De Sanctis, E., Greco, M., Piccolo, M. and Tazzari, S., eds.), p. 929. Bologna: Editrice Compositori 1988

Sau 87 Sauer, P.U.
Few-Body Systems Suppl. 2, 215 (1987)

Saw 64 Sawicki, J.
Nuovo Cim. 2, 196 (1964)

ScA 89a Schmitt, K.-M. and Arenhövel, H.
Few-Body Systems 6, 117 (1989)

ScA 89b Schmitt, K.-M. and Arenhövel, H.
Few-Body Systems 7, 95 (1989)

Sch 50 Schiff, L.I.
Phys. Rev. 78, 733 (1950)

Sch 51 Schiff, L.I.
Phys. Rev. 83, 252 (1951)

Sch 81 Schwille, W.J.
Proc. Int. Symposium on Lepton and Photon Interactions at High Energies (Pfeil, W., ed.), p. 334. Bonn: Physik. Institut, University of Bonn 1981

Sch 86 Schmitt, K.-M.
Doctoral thesis, University Mainz, Mainz (1986)

ScS 81 Schulze, M.E., Saylor, P.P. and Goloskie, R.
Phys. Rev. C24, 2435 (1981)

ScW 90 Schmitt, K.-M., Wilhelm, P., Arenhövel, H., Cambi, A., Mosconi, B. and Ricci, P.
Phys. Rev. 41, 841 (1990)

SeB 85 Seltzer, S.M. and Berger, M.J.
Nucl. Instr. Meth. B12, 95 (1985)

SeB 86 Seltzer, S.M. and Berger, M.J.
Atom. Dat. Nucl. Dat. Tab. 35, 346 (1986)

Sie 37 Siegert, A.J.F.
Phys. Rev. 52, 787 (1937)

Sko 64 Skolnick, M.H.
Phys. Rev. 136B, 1493 (1964)

SkS 74 Skopik, D.M., Shin, Y.M., Phenneger, M.C. and Murphy, J.J.
Phys. Rev. C9, 531 (1974)

SmB 87 Smit, F.D. and Brooks, F.D.
Nucl. Phys. A465, 429 (1987)

SmT 77 Smirnov, S.A. and Trubnikov, S.V.
Teor. Mat. Fiz. 30, 28 (1977) (Sov. J. Theor. Math. Phys. 30, 17 (1977))

SnB 50 Snell, A.H., Barker, E.C. and Sternberg, R.L.
Phys. Rev. 80, 637 (1950)

SoC 69 Sober, D.I., Cassel, D.G., Sadoff, A.J., Chen, K.W. and Crean, P.A.
Phys. Rev. Lett. 22, 430 (1969)

SoK 87 Soderstrum, J.P. and Knutson, L.D.
Phys. Rev. C35, 1246 (1987)

Sor 85 Sorokin, P.V.
Proc. 9th Europ. Conference on Few-Body Problems in Physics (Faddeev, L.D. and Kopaleishvili, T.I., eds.) p. 440. Singapore: World Scientific, 1985

StH 87 Stephenson, K.E., Holt, R.J., McKeown, R.D. and Specht, J.R.
Phys. Rev. C35, 2023 (1987)

StK 85 Stiehler, T., Kühn, B., Möller, K., Mösner, J., Neubert, W., Pilz, W. and Schmidt, G.
Phys. Lett. 151B, 185 (1985)

Str 64 Stranahan, G.
Phys. Rev. 135B, 953 (1964)

TaG 72 Tartakovskii, V.K. and Gurin, Y.L.
Ukr. Fiz. Zh. 17, 1125 (1972)

TaM 60 Taylor, R.E. and Mozley, R.F.
Phys. Rev. 117, 835 (1960)

TaO 89 Tanabe, H. and Ohta, K.
Phys. Rev. C40, 1905 (1989)

TaP 58 Tatro, C.A., Palfrey, T.A., Jr., Whaley, R.M. and Haxby, R.O.
Phys. Rev. 112, 932 (1958)

ThF 86 Thorlacius, A.E. and Fearing, H.W.
Phys. Rev. C33, 1830 (1986)

Tsa 65 Tsai, Y.S.
Phys. Rev. B137, 730 (1965)

Tsa 74 Tsai, Y.S.
Rev. Mod. Phys. 45, 815 (1974)

Tud 67 Tudorič-Ghemo, J.
Nucl. Phys. A92, 233 (1967)

VeK 81 Vesna, V.A., Kolomensky, E.A., Kopeliovich, V.B., Lobashev, V.M., Nazarenko, V.A., Pirozhkov, A.N. and Shulgina, E.V.
Nucl. Phys. A352, 181 (1981)

VFS 89 von Ferber, Chr., Sandhas, W. and Haberzettl, H.
Phys. Rev. C40, 1923 (1989)

VFT 84 van Faassen, E.E. and Tjon, J.A.
Phys. Rev. C30, 285 (1984)

Vil 47 Villars, F.
Helv. Phys. Acta 20, 476 (1947)

VnG 86 Vnukov, I.E., Glavanakov, I.V., Krechetov, Yu.F., Potylitsyn, A.P., Saruev, G.A., Stibunov, V.N. and Tabachenko, A.N.
Pis'ma Zh. Eksp. Teor. Fiz. 43, 510 (1986) (Sov. Phys. JETP Lett. 43, 659 (1986))

VnG 88 Vnukov, I.E., Glavanakov, I.V., Krechetov, Yu.F., Potylitsyn, A.P., Saruev, G.A., Stibunov, V.N. and Tabachenko, A.N.
Yad. Fiz. 47, 913 (1988) (Sov. J. Nucl. Phys. 47, 581 (1988))

VyR 82 Vyas, R. and Rustgi, M.L.
Phys. Rev. C26, 1399 (1982)

Wal 89 Wallace, P.A.
Ph.D. thesis, University of Glasgow, Glasgow (1989)

Wat 54 Watson, K.M.
Phys. Rev. 95, 228 (1954)

Way 37 Way, K.
Phys. Rev. 51, 552 (1937)

WäY 51 Wäffler, H. and Younis, S.
Helv. Phys. Acta. 24, 483 (1951)

WeA 78 Weber, H.J. and Arenhövel, H.
Phys. Rep. 36, 279 (1978)

WhH 58 Whetstone, A.L. and Halpern, J.
Phys. Rev. 109, 2072 (1958)

Wil 56 Wilson, R.R.
Phys. Rev. 104, 218 (1956)

WiL 88 Wilhelm, P., Leidemann, W. and Arenhövel, H.
Few-Body Systems 3, 111 (1988)

WiS 84 Wiringa, R.B., Smith, R.A. and Ainsworth, T.L.
Phys. Rev. C29, 1207 (1984)

Wyn 71 Wynn, W.M.
PhD thesis, Georgia Institute of Technology (1971)

YiH 88 Ying, S., Henley, E.M. and Miller, G.A.
Phys. Rev. C38, 1584 (1988)

Zac 56 Zachariasen, F.
Phys. Rev. 101, 371 (1956)

ZiA 61 Zickendraht, W., Andrews, D.J. and Rustgi, M.L.
Phys. Rev. Lett. 7, 252L (1961)

ZiG 86 Zieger, A., Grewer, P. and Ziegler, B.
Few-Body Systems 1, 135 (1986)

ZiR 79 Zickendraht, W., Rustgi, M.L. and Brandt, R.G.
Nuovo Cim. 52A, 247 (1979)